Walking Scotland's Lost Railways

Walking Scotland's Lost Railways

Track Beds Rediscovered

Robin Howie & John McGregor

Whittles Publishing

Published by
Whittles Publishing Ltd.,
Dunbeath,
Caithness, KW6 6EG,
Scotland, UK

www.whittlespublishing.com

ISBN 978-184995-403-7

DISCLAIMER

The Authors have made every effort to ensure that the information contained in the track bed walks has been brought up to date. However, there is always constant change, whether to track beds, embankments or cuttings. Accordingly, the authors would welcome information of any discrepancies and/or comments from readers. In addition, it should be noted that the ease in traversing overgrown track beds will depend on the time of year. Some walks have been noted as "best done in winter".

Printed and bound in Great Britain by Severn, Gloucester

Contents

Acknowledgements

Valuable advice came fom several quarters, but particular thanks are due to Ewan Crawford and John Furnevel, whose work includes so many telling images of "railway remains". For a sufficient coverage of the Borders and Fife, Clackmannan & Kinross I am indebted to Bill Lynn, who generously allowed unlimited resort to the R.W.Lynn Collection. For the South Central Highlands, there was friendly assistance from Doug Carmichael (Friends of the West Highland Lines) and Dave Ferguson, who made available his own images besides those of L.A.C.Horne and G. Maxwell. I have drawn too on the John Alsop Collection and the Lens of Sutton Archive. For access to the latter, I had the help of David King (North British Railway Study Group).

John McGregor, June 2019

The book would not have achieved fruition without the assistance and co-operation of so many friends and colleagues. From the very start Margaret has been of constant support, especially when I have been away for days on end tramping many a track bed mile. I am also grateful to those friends who accompanied me on some of those track bed walks, including Rhona Fraser, Jim Wyllie, Douglas Paul, Rae Montgomery, Derek Chambers and Judy and Goff Cantley.

Every single day along the way, the folk I stopped and chatted to were friendly, welcoming and glad to offer any assistance....and a mine of information. Many had fascinating recollections, direct or handed down, of railway days; so much so that at times I almost had to tear myself away to get on with the walk. On numerous occasions I was invited into their homes for refreshment. My thanks go to them all.

Maps are an integral part of this book. We are indebted to the ever patient Mick Ashworth of Ashworth Maps and Interpretation Ltd, Bishopbriggs for the design of all the hand crafted maps. These were created by combining current map data with information on historical routes and station sites from old Ordnance Survey one-inch maps.

Robin Howie, June 2019

Scottish Railway Network c. 1914

Line in operation 1914

Walk routes

- South Central Highlands
- Fife
- Borders

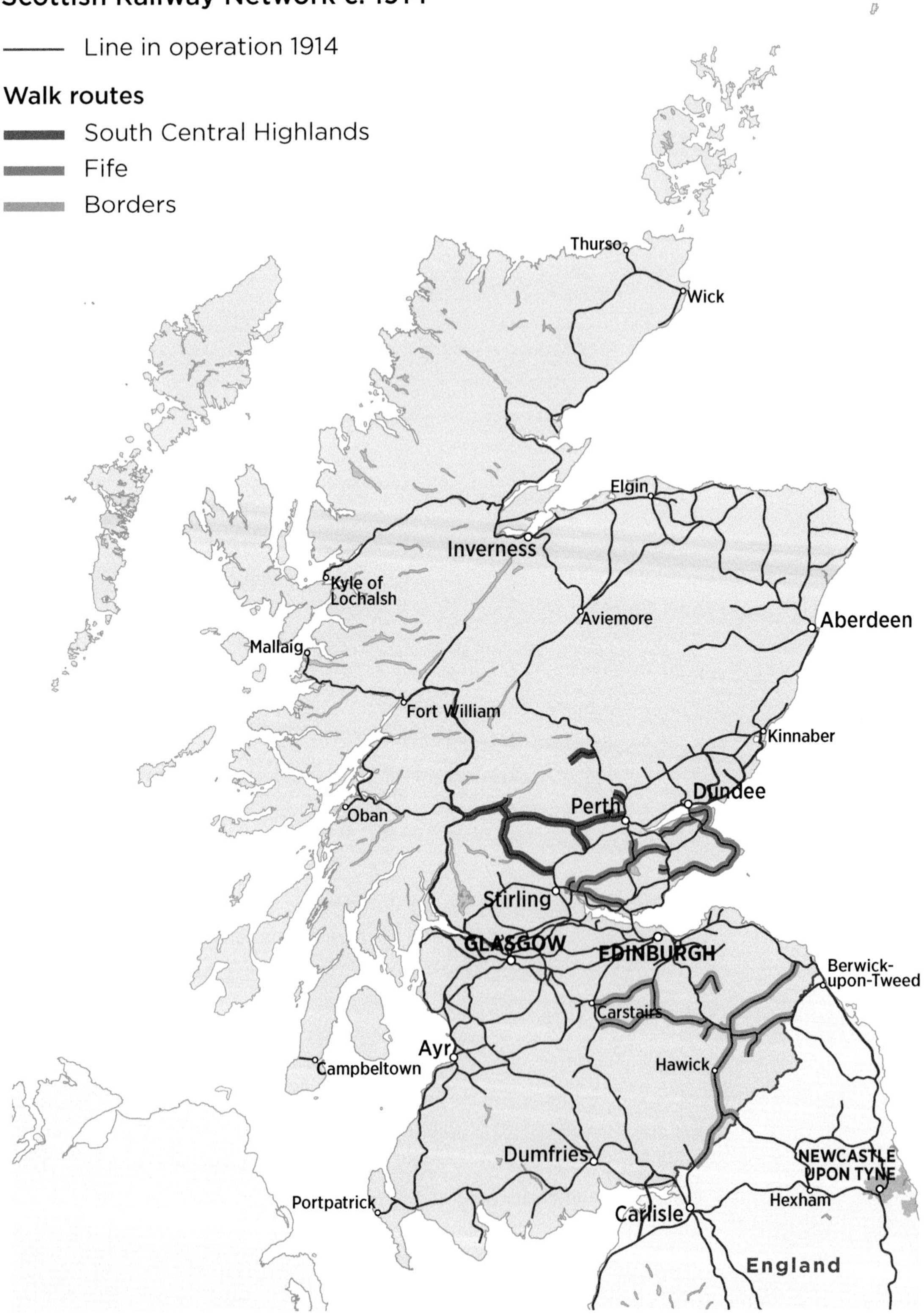

Introduction

The Railway Age – an overview

The idea of a late 18th century-early 19th century industrial revolution, subsuming parallel "revolutions" in agriculture, population, transport and urban growth, is firmly fixed in popular culture. Dramatic though this period was, historians are at pains to downplay the notion of an abrupt and decisive break with the past: they prefer to deal in terms of an accelerating rate of technological change, on a relatively long time-scale. After the earliest successful production of machine-spun thread, to develop a broad textile industry of factory spinning and weaving required some 120 years. Until c.1850 rural decline was relative rather than absolute (save in the special circumstances of Highland Clearances and Irish Famine). At the middle of the 19th century only just over 50% of the population lived in a more-or-less urban environment, while agriculture and domestic service remained the largest categories of employment. Purchasing power (for non-necessities) and availability of capital still depended to a measureable extent on the cost of food and fodder, and thus on the quality of each harvest – which partially explains the jagged boom-slump pattern of the 1830s and 1840s.

To quibble over the expression "railway revolution" is pedantic but here too the story is more *evolutionary* – from 18th century wagon ways and horse traction, as adjuncts of the coal industry, to the first inter-city lines and steam locomotives c. 1830. Brittle cast-iron rails were vulnerable, and primitive suspension as well as low boiler pressure limited the efficiency of the early "travelling engine". Progress lay in the gradual advance of iron technology. With a good deal of trial and error, only from the 1860s did modern track become the general rule (heavy wrought-iron rails, keyed in cast-iron chairs; fish-plated rail joints; and robust but elastic and well-drained ballast). Firebox improvements were needed before locomotives could consume raw coal rather than coke. Steel boilers, rails and tyres became general only in the 1870s and 1880s. The sequence "cast iron, then wrought iron, then steel" also governed the scope of civil engineering, which was intimately linked with the expansion of the railway network. (How the water breaks of Forth and Tay were overcome is a story in point.)

Railway development echoed, at a faster pace, the development of canals – a first stage when relatively local schemes were financed by those with an immediate need for improved transport; a second stage when *transport in itself* had come to seem a rewarding investment, so that more ambitious, more speculative ventures multiplied. If we are to speak of Britain's *transport* revolution, the railways were, it can be argued, not fundamental. Road, river and harbour improvement, together with canals, had brought a useful degree of economic integration long before 1800. The railways *completed* a process already well-advanced (cf. the USA where the railroads proved essential to economic and political unification, continent-wide). The central role of railways in 1830s-1860s Britain was to broaden and deepen the narrow and still vulnerable industrial economy, while the surge of railway investment, despite the hazards of bubble and bust in the 1840s, ultimately expanded and strengthened the country's financial institutions.

It follows that Britain's railway companies in their years of dominance were identified with free trade and private enterprise, which by c.1850 had won an ideological victory over the traditional precepts of mercantilism and protection. But there was a contradiction. Despite vigorous inter-company competition, the railway industry *as a whole* soon achieved a monopolistic position and an influential voice (the so-called Railway Interest) inside Parliament. The railways, by their very nature, became early examples of big business, and this was confirmed by amalgamations and take-overs. That they were over-powerful and ought to be more closely regulated was a recurring cry. Though government direction or overview of railway projects, leading to state ownership, would become the normal European pattern, this approach had been rejected in Britain. (It *was* applied in British India.) State purchase of established companies, theoretically an option, ceased to be practical politics for the

NBR Selkirk branch train *Alsop*

rest of the 19th century. A classic laissez-faire argument was much quoted on the railways' behalf. Exemplifying "private risk for public benefit", they deserved a wide immunity from interfering legislation. Though this ceased to be a tenable position, government regulation advanced slowly, even in matters of safety.

By the 1880s, however, traffic pooling and statutory rates had very much reduced open competition. Covert competition continued – in general facilities, freight rebates and quality of passenger service. In the long term the railways faced increasing costs, labour troubles and new legislative requirements – and the companies increasingly looked for compensating concessions. Would the state free the railways from "common carrier" obligations, encourage rationalisation and permit a new round of amalgamations? For the Liberal Government of 1905-14, beset with constitutional crises and incipient party splits, the railways' future was one more intractable problem. Had the 1914-18 War not broken out, they or their successors would have been obliged to find a new formula.

In the event an answer came in the shape of emergency wartime Control and post-1918 De-Control, when the companies were returned to private ownership but compulsorily combined into four large Groups (1923), as the alternative to outright nationalisation. During the 1920s and 1930s the "Big Four" secured a measure of indirect subsidy, including exemption from local authority rates at Westminster's expense.

Scottish railways

From the 1860s, though small local companies, often penurious, endured, much of the nation's route mileage came to be controlled by fewer than twenty large units. Five among them were Scottish – the **North British**, which absorbed the Edinburgh, Perth & Dundee and the Edinburgh & Glasgow; the **Caledonian**, which absorbed the Scottish Central and the Scottish North Eastern; the **Highland,** in which all the early Inverness-based companies were united; the **Glasgow & South Western**, which grew out of the Glasgow, Paisley, Kilmarnock & Ayr; and the **Great North of Scotland**, which remained a compact system based on Aberdeen. (The five are listed by route mileage. The Great North by 1914 possessed some 330 route miles, where the North British had more than 1,000.)

North British-Caledonian enmity was a given of Scottish railway politics from the 1870s into the early 20th century, and this helped to preserve the independence of Highland and Great North, whose through traffic was sought by both protagonists. After a barren contest to acquire the Glasgow & South Western (1889-91), North British and Caledonian found their way to a grudging but durable détente. Great North of Scotland and Highland had a contested frontier, exchanging traffic (more-or-less unhappily) at Elgin, Keith and Boat of Garten. Caledonian and North British were enmeshed, despite their mutual enmity. For entry to Aberdeen and connection with the Great North of Scotland, the North British exercised running powers over the Caledonian north of Kinnaber (Montrose). Caledonian domestic traffic north from Edinburgh depended on running powers over the North British through Falkirk to Larbert.

West Coast Anglo-Scottish traffic was in Caledonian hands north from Carlisle. In the competing East Coast alliance, the North British was something of a sleeping partner. Until Forth and Tay were both finally bridged (1890), all English through traffic to and from the north of Scotland ran via Caledonian Stirling. Moreover, the (English) North Eastern Company brought the principal cross-Border trains through Berwick into Edinburgh, while a reciprocal but scarcely even-handed agreement gave the North British entry to Newcastle via the Border Counties Railway and Hexham. In 1876 a third Anglo-Scottish partnership came into being when the Midland Railway reached Carlisle, to ally with the Glasgow & South Western Company and the North British. The latter's Waverley Route by Hawick and Galashiels carried the Midland Company's traffic for Edinburgh and the North.

Several short but strategic lines were placed, sooner or later, under joint ownership. Glasgow-Paisley and Glasgow-Kilmarnock were shared by the Caledonian and the Glasgow & South Western; Dundee-Arbroath and Dumbarton-Balloch (Loch Lomond) were shared by the Caledonian and the North British. The Forth Bridge Railway (Dalmeny-Inverkeithing) was jointly owned by the North British and all three of their English allies – the Great Northern, the North Eastern and the Midland each contributed to the cost of the Forth Bridge. The Portpatrick & Wigtownshire Railway, a larger entity, became the shared property of the Caledonian, the Glasgow & South Western, the London & North Western and the Midland. The two Scottish partners at first worked the line alternately. All these arrangements were essentially inter-company treaties. Without overall government direction on the Continental model, the British railway network was overwhelmingly the product of private enterprise and authorised project by project.

By the 1860s a pattern of core trunk lines had been laid down and further additions to the network, more or less desirable, were not infrequently initiated by engineers and contractors, who exploited both the ambitions and the fears of established companies. Set-piece railway battles in Parliament, greatly adding to first costs, remained a British characteristic. The end result was overprovision in certain areas and gaps elsewhere – especially in the Highlands. The Stirling-Perth-Inverness trunk line was completed by 1863 and the Further North route, beyond Inverness, to Wick and Thurso by 1874. Two cross-country lines stretched to the west coast – the Dingwall & Skye, opened in 1870 (to Strome Ferry) and the Callander & Oban, completed in 1880. Late in the day the West Highland Railway came to link the Central Belt with Fort William (1894) and Mallaig (1901). Having absorbed the Dingwall & Skye in 1880, the Highland also gathered in the nominally independent "Further North" companies (1884). The Callander & Oban, though a Caledonian client, retained a more genuine independence until the 1923

Riccarton Junction: NBR local train in south bay; veteran ex-Edinburgh & Glasgow Railway 2-2-2, 1005 *Lynn*

Grouping. The West Highland, though formally separate until 1908, was from the first "the North British by another name".

Here government had played some part. Distress and unrest in the Highlands and Islands made action imperative, whereupon the Napier enquiry (1883-4) and then the Lothian enquiry (1889-90) recommended transport improvements – including light railways (never to be achieved) for Lewis and Skye. Controversial state subsidy saw the Dingwall & Skye extended to Kyle of Lochalsh (1897) and the West Highland to Mallaig. Nevertheless the railway map remained disjointed. The Highland Company's Aviemore cut-off (1898) made sure that the Perth route would remain the rail artery of the North, while the three wandering lines to the west coast continued to serve their three distinct railheads (Oban, Mallaig and Kyle). The Invergarry & Fort Augustus Railway, though aiming at a West Highland-cum-Great Glen route to Inverness, was fated to remain a West Highland appendage, purchased by the North British in 1914. The Ballachulish branch (Callander & Oban) was not extended to Fort William.

By the 1890s and 1900s the Scottish companies were increasingly intent on economy and retrenchment. Among signs of the times were the comprehensive Clyde Coast rail-and-steamer traffic pool, established by North British, Caledonian and Glasgow & South Western, and the attempted merger (1906) of Highland and Great North of Scotland. Respect for one another's preserves became the unadmitted reality, while new schemes, which had already much declined in number, dwindled away. Without a clear lead from central government, there was little to encourage new railway projects, useful or otherwise. Unrealistic hopes attended the Light Railways Act of 1896, which shifted the burden of assistance to local authorities. These seldom proved sufficiently pro-active. Worthy light promotions needed only a Board of Trade Order; but the *procedure* was still cumbersome and in Scotland only a handful of such schemes came to fruition. The motor bus and motor lorry were already recognisable as long-term competitors, and the 1914-18 War would emphasise their potential.

A private-enterprise railway system was at the same time a state asset. What should be done when, on simple commercial grounds, a branch or little-used line faced closure? This question did not loom seriously until the inter-War years – but the Invergarry & Fort Augustus provided a notorious early example, rescued from abandonment in 1913 by the intervention of Inverness-shire County Council. The contractions of the 1920s and 1930s, with a further round in the first years of Nationalisation, were hard to contest given the success and (usually) the popularity of competing road services. (Goods business, with minimal maintenance and basic facilities, often survived the withdrawal of passenger workings; a weekly coal train, or seasonal traffic, might well remain worthwhile.) Altogether, these disappearances dented the railway map which the 19th century had bequeathed; but closures on a drastic scale waited for the Beeching Plan implemented in the 1960s.

Under the Grouping the North British and the Great North would become **LNER** (London & North Eastern Railway), while the Caledonian, the Highland and the Glasgow & South Western would become **LMS** (London, Midland & Scottish Railway) – a division which perpetuated some traditional rivalries. Proposals for an all-Scotland Group, rejected after 1918, would find favour after a fashion thirty years later. Take-over by the state during the 1939-45 War led on to nationalisation (1948) and the creation of **British Railways, Scottish Region.** An LNER: LMS divide, with echoes of older rivalries, persisted under public ownership, at least as a cast-of-mind, as did a conservative attitude to operation and management. Government "short-termism" did the railway industry no favours – commitment to one last generation of standardised steam locomotives was overtaken within a decade by a stop-go programme of electrification and an ill-considered rush to dieselise the rest of the network.

Whatever the merits of privatisation in the 1990s, today's railway renaissance has surer foundations and shows more staying power.

Public Access

Scotland's outdoors provide great opportunities for open-air recreation and education, with great benefits for people's enjoyment, and their health and well-being. The Land Reform (Scotland) Act 2003 ensures everyone has statutory access rights to most of Scotland's outdoors, if these rights are exercised responsibly, with respect for people's privacy, safety and livelihoods, and for Scotland's environment. Equally, land managers have to manage their land and water responsibly in relation to access rights.

The **Scottish Outdoor Access Code**, which can be found, and downloaded, at www.outdooraccess-scotland.com, provides detailed guidance on these responsibilities. The Code provides a practical guide to help everyone make informed decisions about what best to do in everyday situations.

The Code is based on three key principles:

- **Respect the interests of other people.**
  Acting with courtesy, consideration and awareness is very important. If you are exercising access rights, make sure that you respect the privacy, safety and livelihoods of those living or working in the outdoors, and the needs of other people enjoying the outdoors. If you are a land manager, respect people's use of the outdoors and their need for a safe and enjoyable visit.

- **Care for the environment.**
  If you are exercising access rights, look after the places you visit and enjoy, and leave the land as you find it. If you are a land manager, help maintain the natural and cultural features which make the outdoors attractive to visit and enjoy.

- **Take responsibility for your own actions.**
  If you are exercising access rights, remember that the outdoors cannot be made risk-free and act with care at all times for your own safety and that of others. If you are a land manager, act with care at all times for people's safety.

With access rights come responsibilities.
Track bed walkers, with or without dogs, should take particular care in traversing pasture land. Cattle, invariably inquisitive, may be encountered. Cows are extremely protective of young calves, and sheep, especially during the lambing season, can be disturbed. Cows (never mind bulls) may react aggressively to the presence of dogs even when on lead. Irresponsible dog walking can cause problems for land managers. If there is any doubt about access, make enquiries or undertake appropriate detours. Leaving your dog at home prior to a pasture traverse is the prudent option.

Electric fences may well be encountered. Where an electric fence crosses a public pathway, a non-electrified gate or a stile should be provided. However, where the track bed bisects farm land used by cattle, safe crossing points may be lacking.

When (carefully) straddling electric wires, lengths of old bike tyres may provide the necessary insulation. Even better is the Montgomery Fence Straddler....a strong plastic tube, or a wider section of guttering, cut lengthwise and lined with duck-tape. A foam or towel backing will give more comfort and also cope with barbed wire.

The Railway Legacy

Although walkers may be unaware of the tangled history of the development of the railway system during the Victorian era, they will have heard of, or experienced, the drastic 1960s Beeching axe. However, in more recent times Scotland has experienced a railway revival – principally in the Greater Glasgow area but with new stations and station re-openings elsewhere. The Borders Railway (2015) is something on a very different scale.

Nevertheless, Scotland still has hundreds of miles of "dismantled railways", the term used by the Ordnance Survey, and the track beds give scope for many a walk. Some track beds have been "saved" as tar macadam walkway/cycleway routes; some have become well-trodden local walks. The remainder (the majority) range in walking quality from good, to overgrown, to well-nigh impassable.

In every project the railway engineer strove to balance "cut" and "fill" and set a "ruling gradient". This has left a general legacy of often impressive cuttings and embankments – and steady, sometimes imperceptible, ascents and descents from the walker's perspective. For railway passengers (and crew), this was only noticed to the extent that distant views came and went. For today`s track bed walker, cuttings and embankments take on a different aspect.

Cuttings frequently tend to be wet and overgrown. Trees and scrub have flourished. Where farmers have recovered railway ground, agricultural equipment or stores of fodder may bar the way. A dry and clear cutting is a thing to be welcomed! Embankments on the other hand are invariably dry and, on the whole, less overgrown, giving views of the surrounding landscape. A study of the Ordnance Survey map before setting out will give a flavour of what to expect.

Newcastleton: ex-NBR Class J36 0-6-0, carrying BR number 65232, prepares to bank a goods train north to Whitrope *Lynn*

In the earliest days of railway construction, the slopes of cuttings and embankments were generously pitched. The prudent limits of safety once established, steeper sides became acceptable; but, abandoned over decades, these became more vulnerable to erosion. Good drainage was essential but many of the line-side ditches have not survived so that cuttings lie wet or are scoured by the winter spates that invade the track bed. The risk of line-side fires was much reduced when diesel and electric traction replaced steam–but so was the traditional routine of trimming back trees and vegetation.

Conditions under foot and time-of-year may dictate detours (all described). However, the intrepid track bed explorer can ignore some of these. For example, by the edge of Loch Lubnaig, the track bed in parts has been lost or washed away, yet persistence is rewarded in reaching later sections that offer secluded and easy strolls. This is what track bed exploration can be like!

A summary railway history introduces every chapter of this essentially "walking" book. However, the walks are treated end-to-end as latter-day voyages of exploration, though the original railway route may have come into being incrementally.

Borders

The Borders

The North British Railway eventually dominated south-eastern Scotland (broadly the Edinburgh-Berwick-Carlisle triangle), gaining outposts at Port Carlisle and Silloth (on the Solway's English shore) and penetrating to Hexham, Morpeth and Rothbury, deep in Northumberland. However, the Caledonian Railway was from the outset a formidable neighbour, while the (English) North Eastern Railway, formed by amalgamation in 1854, became a sometimes overbearing partner. The North British advance from Teviotdale (Hawick) into England, by Whitrope and Liddesdale, was contested; and at first the Edinburgh-Hawick-Carlisle line had little more than local significance – Anglo-Scottish traffic through Carlisle was controlled by the Caledonian Company and their West Coast partner, the London & North Western. The designation Waverley Route became generally recognised only after 1876 when the Midland Railway completed their Settle & Carlisle line and found Scottish allies – the North British for through traffic to Edinburgh and the North, the Glasgow & South Western for connection with Ayrshire and Clydeside via Dumfries.

Walter Scott, seeking a title with no unwanted connotations for his 1814 novel, chose "Waverley" (in Surrey, today a local government district with borough status). By virtue of the book's success, "Waverley" came to denote his entire series of historical fictions. *Sir* Walter from 1820, he died in 1832. Edinburgh duly acquired a Scott Monument and a Waverley Bridge, which bestrode the western end of the city's General Station (so styled by c.1850). Waverley Bridge Station and then simply Waverley became the preferred and ultimately official usage (cf. how London's Waterloo Bridge Station became just Waterloo.) Only then did the North British Company realise the resonance of "Waverley" and its potential for publicity. To the East Coast Route (by comparison a hum-drum label) the North British contributed little; but between Edinburgh and Carlisle the Midland Company's through traffic was entirely in North British hands and in charge of their latest locomotives. (For

The BR name board reads 'Change for Innerleithen and Peebles'; Class A4 4-6-2 'Merlin' on main line train *Lynn*

good measure two successive express engines were named "Waverley", as was one of their Clyde Coast paddle steamers, in old age a casualty at Dunkirk.)

Though the North British Company emerged as Scotland's largest (by route miles), after the amalgamations of the 1860s, their system long remained ramshackle. In 1871 surrender to the Caledonian (likewise enlarged by amalgamation) was only narrowly averted. From 1869, moreover, the North Eastern Railway exercised running powers into Edinburgh, handling all East Coast Anglo-Scottish passenger traffic. Salvation lay in the eventual unification of the North British network by the great bridges over Forth and Tay, and also in their Midland partnership (despite the consequent friction with their earlier East Coast associates). Confidence grew when the "Bridges Route" at last became available for through traffic, both domestic and Anglo-Scottish, north from Edinburgh – to Perth and Inverness, and to Dundee and Aberdeen. In assertive mood, the reinvigorated North British attempted to reclaim their original Edinburgh-Berwick main line from North Eastern occupation (1897). In the end the latter's rights were upheld, though with face-saving concessions. From 1923 the East Coast route would become LNER throughout, while London-Carlisle-Edinburgh services via the Waverley Route required an LMS-LNER arrangement.

The North British story properly began in 1844 when construction to Berwick was authorised (opening came two years later). The new company quickly purchased the Edinburgh & Dalkeith Railway (dating from 1826), as the stem of an early thrust southward into Tweeddale and Teviotdale. The Edinburgh & Hawick Railway, approved in 1845 and quickly vested in the North British, was completed in 1849. The Border Union Railway won its

Act in 1859, taking the route on by Liddesdale to Carlisle (1862); but the immediate rewards were slim. Teviot-to-Esk was arguably the better option, serving Langholm directly; instead, the town would obtain a branch from Ridings Junction, (opened in 1864). There were those who, questioning whether the line should be built at all, urged an accommodation with the Caledonian (by which means a little English traffic might be gained). It was a serious fear, too, that a Caledonian blocking scheme would penetrate to Langholm and perhaps onwards to Hawick. At the Border Union's southern end, Caledonian obstruction was overcome by an arrangement with the Port Carlisle Railway, which the North British leased (1862) and eventually absorbed (1880).

The Kelso line, authorised in 1846 and diverging at St Boswells, opened in 1850. It reached initially to Wallace Nick. Meanwhile the complementary branch from Tweedmouth, on the York, Newcastle & Berwick Railway (a constituent of the North Eastern Company), had halted at Sprouston (1849). In 1851 landowner opposition was overcome and the gap closed, with Kelso's station sited across the Tweed, at Maxwellhaugh. Thus Edinburgh-St Boswells-Tweedmouth became a secondary through route, in part double-tracked. Maxwellhaugh-Sprouston was at first jointly owned; but Kelso was the logical terminus of the North Eastern branch, and this provided the only example of an English railway company working into Scotland. (Coldstream, however, had its station on the "wrong side of Tweed".) Also authorised in 1846 was the Jedburgh branch, from Roxburgh Junction. Construction, after a delay of several years, was undertaken by the nominally independent Jedburgh Company (1855), acquired by the North British in 1860. Double track was provided, against the possibility of a through route beyond Jedburgh and over Carter Bar into England (below).

Approved in 1855, the Peebles Railway ran from Hardengreen Junction via Leadburn and Eddleston. Completed in 1859, it was leased by the North British, who did not acquire the line outright until 1876. In the interval this constituted a temptation for the Caledonian, who proposed a connecting line, from Carstairs to Leadburn via Dolphinton. And by means of the Symington, Biggar & Broughton Railway (completed in 1860) and its Peebles extension (1864), the Caledonian did indeed win through to Tweeddale. To repel any further Caledonian probes, the North British backed and subsequently absorbed both the Leadburn, Linton & Dolphinton Railway (opened in 1864) and the Galashiels & Peebles Railway (completed throughout by 1866). Parliament approved in principle (the precise route was detailed later) a Caledonian branch from Carstairs via Dunsyre to Dolphinton, for end-on connection with the Leadburn-Dolphinton line, which was achieved in 1867; but both parties thereafter actively *discouraged* through traffic and Dolphinton remained de facto a Caledonian: North British frontier. At Peebles the link across the Tweed, from Caledonian station to North British station, lacked the promised triangular junction whereby the Caledonian might have pushed on, under running powers or joint ownership, to meet the Waverley Route at Galashiels. In effect Peebles became another frontier. The mutual suspicions of Caledonian and North British died hard. Some thirty years later the North British would entertain exaggerated fears that their old enemy would find fresh opportunities in the light railway south from Broughton, built to service Talla reservoir and worked by the Caledonian. (Despite hopes that the line would endure, it was sold and dismantled in 1910).

The Selkirk & Galashiels Railway was a short uncomplicated branch – authorised in 1854, opened in 1856 and taken over by the North British in 1859. The Berwickshire Railway, authorised in 1862, extended the existing Duns branch (which diverged from the Edinburgh-Berwick main line at Reston) first to Earlston (1863) then to Ravenswood Junction, just north of St Boswells, (1865). Like the Peebles Railway, it was brought in to the North British fold in 1876. The completed Reston-St Boswells line paralleled and in some degree duplicated the Tweedmouth-Kelso-St Boswells link. North British route mileage (which topped 1,100 at its pre-1914 maximum) was inflated by such over-provision (cf. the leaner, more coherent Caledonian network). Their recurring financial difficulties are partly explained by the burden of so many dubiously worthwhile lesser lines (not just in the Borders). In mid-19th century Britain established railway companies frequently offered monetary contributions, working agreements and dividend guarantees to new projects, so as to enlarge or defend their own territories; but the North British proved imprudently lavish. Blocking tactics of course played a part, and strategic calculation was frequently overtaken by events, turning grand vision into barely profitable branch. The public gained little. By the 1860s piecemeal construction (as sketched above) had created a continuous route of

Tank engine modified for Lauder Light Railway *Lynn*

sorts from lower Tweed to upper Clyde; but through traffic was minimal and inhibited by Caledonian-North British enmity.

The Border Counties Railway was first authorised (1854) as a Hexham-Falstone branch. Its ancestors were among the earliest Anglo-Scottish railway schemes, and these included a route from Hexham (on the then Newcastle & Carlisle Railway) to Jedburgh and Melrose, there dividing for Edinburgh and Glasgow. It is unsurprising, therefore, that the Border Counties promoters, indicating an intention to continue their line into Scotland, angled for North British support. Their manoeuvres gave a further twist to the already complicated preliminaries of the Border Union Railway (above), before the North British resolved to capture both companies (1859-60). The Border Counties opened to Falstone in 1861. Extension to Lees Bog (Riccarton Junction) on the Border Union followed (1862), whereby the Border Counties became an adjunct of the Waverley Route. Moreover, the Wansbeck Valley line and its Rothbury branch, diverging from the Border Counties, would be added to the North British empire in Northumberland. These were modest prizes in themselves but revenue was not the first consideration. The North British valued their two "back doors" to Newcastle, via Hexham and at Morpeth, which tempered the power of the North Eastern Railway, immovably posted at Berwick.

The Light Railways Act of 1896 was designed to help fill the remaining gaps in the British network. There were few gaps in the Scottish Borders, but a light branch to Lauder, from Fountainhall on the Waverley Route, won the North British Company's blessing. The requisite Board of Trade Order was obtained in 1898 and the line opened in 1901. "Light" in earlier years had been an imprecise label, which embraced not only lines so conceived but also conventional lines of relatively flimsy construction. The railway to Lauder exemplified the new standard.

Decline began between the Wars. Lauder lost its passenger service in 1932. Leadburn-Dolphinton closed to passengers in 1933. That same year St Boswells-Kelso and Roxburgh Junction-Jedburgh were singled. Carstairs-Dolphinton survived precariously until 1945. Thereafter a managed retreat was the only choice. Jedburgh's passenger trains ceased in 1948 when floods damaged the Teviot viaduct – it was restored sufficiently to sustain a basic freight service until 1964. In 1955 the St Boswells-Berwick passenger timetable (trains reversed at Tweedmouth) was much reduced, and the end came in 1964. Goods traffic between St Boswells and Kelso continued until 1968. Meanwhile the Border Counties Railway had closed in 1958. The floods of 1948 severed the Berwickshire Railway. No trains would run again west from Duns, though a St Boswells-Greenlaw goods service was restored and lasted until 1965. Reston-Duns closed to all to traffic from 1966 – fifteen years after passenger workings ended. The Selkirk branch carried no passengers after 1951. When general goods traffic ceased (1964), a short stub of the abandoned branch survived two years more, serving the Galashiels gas plant. Symington-Peebles closed to passengers in 1950, though freight workings at first continued; curtailed to Broughton in 1954, these ended in 1966. Save for

Gas lamp standard with station name *Lynn*

Hardengreen Junction-Hawthorden (for Penicuik), the secondary loop through Peebles to Galashiels lost all its traffic in 1962.

In an effort to halt the shrinkage of railway business, the LNER had turned to Sentinel steam railcars. Introduced on the Selkirk branch and the Leadburn-Dolphinton line, they met with some success, despite their restricted availability. However, like the diesel multiple units (DMUs) later deployed by British Railways, they could not attach other vehicles. Where passenger bookings continued, alongside a still-worthwhile traffic in livestock, potatoes or domestic coal, locomotive-hauled mixed trains remained the most economical provision. In 1956 the first Modernisation Plan DMUs allocated to Scotland were put to work on the Edinburgh-Peebles-Galashiels route. Initial results were heartening but the slide in passenger numbers soon resumed.

The story of closures in the Borders has its oddities. The Lauder line, which might well have succumbed to the 1948 floods, survived until 1958. At Peebles, the former Caledonian/LMS goods yard remained open until 1959, accessed by the junction spur across the Tweed. (Logically but almost too late, British Railways had named the two stations Peebles East and Peebles West.)

Closure of the Waverley Route to passenger traffic in 1969 felt peculiarly abrupt and brutal. (Limited goods workings survived a little longer.) Though the line had been run down, to the very end it retained its inter-city, Anglo-Scottish trains, which vanished all at once along with local services. As in other regions, contraction of the nationalised railway network had become inexorable, eventually leaving the Edinburgh-Berwick-Carlisle triangle empty. While Galloway, Strathmore and the North East were denuded too, the Borders towns, with a sizeable aggregate population, were especially disadvantaged.

The restoration (2015) of a passenger service south from Edinburgh to Galashiels and Tweedbank, in the shape of the Borders Railway, was indeed a landmark event.

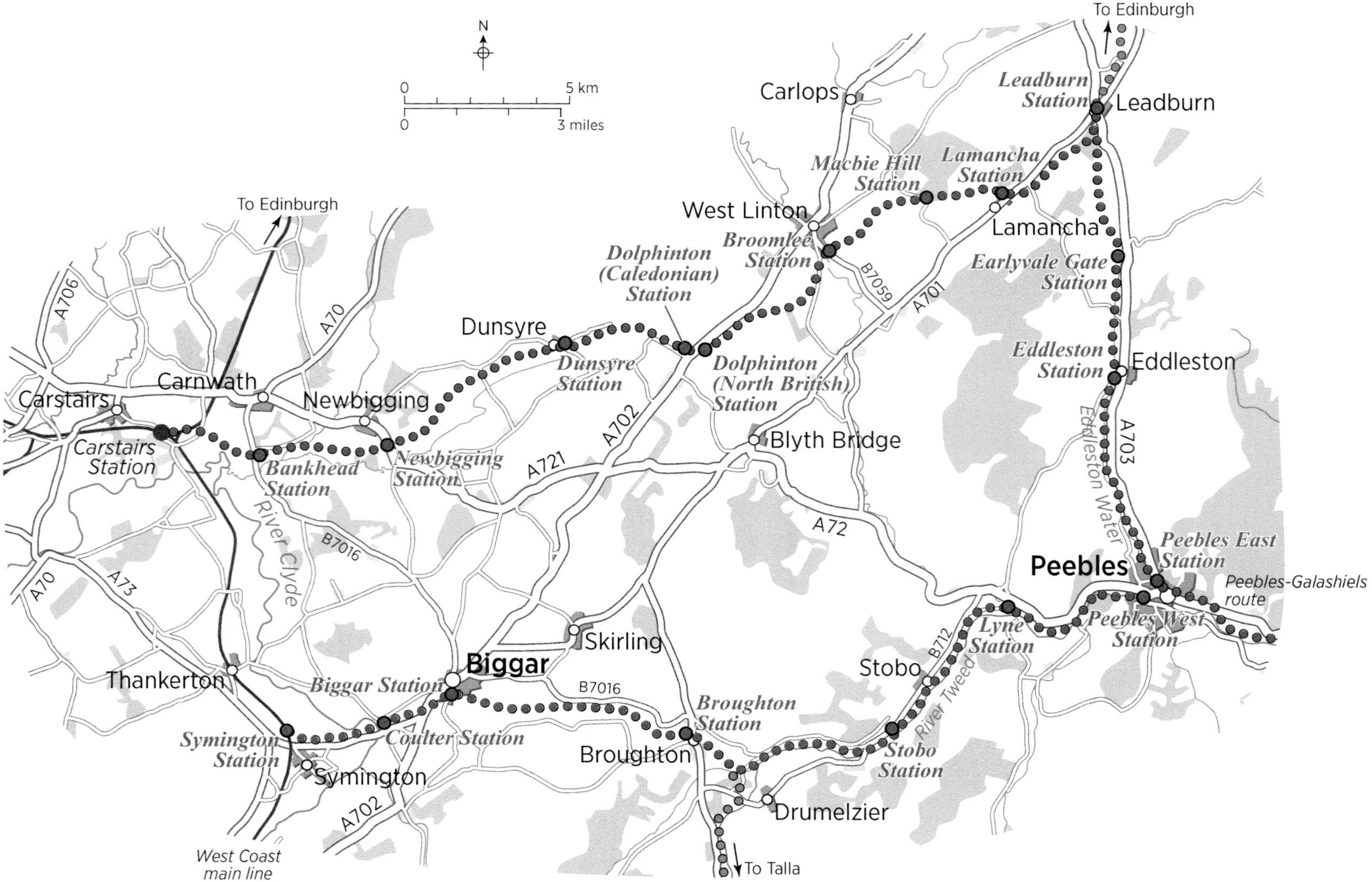

N
0
5 km
0
3 miles
To Edinburgh
Leadburn Station
Leadburn
Carlops
Macbie Hill Station
Lamancha Station
Lamancha
West Linton
Broomlee Station
B7059
A701
Earlyvale Gate Station
Dolphinton (Caledonian) Station
Dunsyre
Dunsyre Station
Dolphinton (North British) Station
Eddleston Station
Eddleston
A703
Eddleston Water
To Edinburgh
A706
A70
Carnwath
Carstairs
Newbigging
Carstairs Station
Bankhead Station
Newbigging Station
A721
A702
Blyth Bridge
A72
River Clyde
B7016
A70
A73
Peebles East Station
Peebles
Peebles-Galashiels route
Peebles West Station
Lyne Station
B712
Skirling
Biggar
Thankerton
Biggar Station
B7016
Stobo
River Tweed
Broughton Station
Symington Station
Coulter Station
Broughton
Stobo Station
Symington
A702
Drumelzier
West Coast main line
To Talla

Leadburn-Peebles

From Hardengreen, the Peebles railway followed the valley of the North Esk, crossing from Midlothian into the Borders at Leadburn.

At the very busy A701/A703 junction, opposite the Moffat A701 sign, a short lane gives access to the embankment on the Midlothian side. Pass by one house and to the left, at the end of the lane, an overgrown grassy area widens out into the site of Leadburn station.

For a time a picnic area but now overgrown, the railway layout (side platform and island) is still identifiable. Leadburn is still discernible on one large notice-board. All buildings have gone, including the high signal box which commanded the junction. The farthest house to the north by the A6094 was the stationmaster`s, now an extended private dwelling.

With intermediate stations at Eddleston and Earlyvale Gate (albeit the latter closed in 1857), the charming 10-mile Leadburn to Peebles track bed, parallel to the Eddleston Water and the A703, can easily be tackled in one day. A regular bus service provides a speedy return. Long standing plans for a cycleway path, similar to that from Peebles to Innerleithen, have yet to come to fruition.

Map 73 is required.

Leadburn Junction, landscaped remains *Furnevel*

Leadburn to Eddleston 5½ miles

Return to the A701/A703 junction. Head south briefly on the A703, noting the road signs... Welcome to the Scottish Borders and Peebles 10 miles. The railway bridge across the road has completely disappeared. Cross the A703 with considerable care. Following a worn way slanting left, climb to the far side embankment from where a winding path leads to the broad grassy track bed. Traverse the Leadburn Community Woodland (refer to the approach to Leadburn from Lamancha) to the site of the junction where the Leadburn, Linton & Dolphinton Railway diverged (refer to Carstairs–Dolphinton-Leadburn Walk). Beyond a track and a parking area, a short overgrown path, clearing the woodland, leads to open moorland.

At first a broad mossy way, the track bed crosses a bridge high above a burn and then begins the gentle descent towards Eddleston. Cross a concrete farm road and follow the field edge to avoid a cutting within a wooded strip. Regain the track bed which gently curves towards Nether Falla. Cross a possibly muddy cattle feeding area (an inelegant detour may be required) to reach an old white-painted cottage. The line of the track bed follows the edge of a large quarry, Cowieslinn, but the bridge over the quarry road has gone and ahead the track bed is lost. Detour time.

Follow the Nether Falla road to the A703 and head south beyond the quarry entrance (on the right) and the road to Temple (on the left). By the latter is an old white-painted sign, Peebles 7 miles.

Later turn right to the pretty hamlet, the well-named Waterheads. Note on the right the embankment and the bridge spanning the Shiplaw Burn, one of many streams that gather to the Eddleston Water and thence to the Tweed. Continue past the house on the right, Nether Linnfall, where there was a level crossing. Mapped as Earlyvale, this is the probable site of the short-lived Earlyvale Gate station.

Turn left at a Y-junction to a timber yard where logs are split to be sold at garages etc for domestic use. At the far end, slant left to regain the attractive track bed which crosses the infant Eddleston Water; the mapped bridge is easily seen from the A703. Though not entirely walkable, the track bed then follows the narrow strip between Water and road all the way to Eddleston. Pass a brick chimney, the remnant of a surfacemen`s hut, and so to a white house on the far side of the road to Shiplaw. Follow the left-side field edge and regain the track bed at a strip of wood beyond the curtilage of the house. Note on the right the flood protection palisade by the eroded banks of the Water. Carefully cross a narrow bridge over a burn to reach Darnhall Mains, whence the track bed to Eddleston has been ploughed up.

Keeping a white house to the left, follow the field edge close by the ever widening and meandering Water. At the end of the field, and the abutments of the bridge over the Water, cross a fence and follow a path that curves left through a plantation to the A703/Old Manse Road junction. Go down the latter to Eddleston, a delightful conservation village now bypassed by the A703. Station Road is lined by charming white-painted cottages and at the immediate corner is the stationmaster's house.

The district is known as Black Barony. An imposing 16th century tower house, Barony Castle, once home to the Scottish Ambulance College and now the Mercure Barony Castle Hotel, is situated in the Fairy Dean, the narrow wooded valley formed by the Fairydean Burn. During World War Two the castle was requisitioned, becoming a Staff College for training Polish officers. Located in the grounds is the B-listed Great Polish Map of Scotland, a three-dimensional outdoor scale model in concrete, claimed to be the world`s largest.

Continue past the entrance to the hotel and turn left on Station Lye; the site of the station, amid houses and gardens. The station building and booking office, with two authentic railway lanterns outside, are readily identified.

Eddleston to Peebles East station 4½ miles

Return from Station Lye, turn left and enter a play area and football pitch overlooked by a cemetery. Cross a stile at the far end and, by following the field edge for a short distance, regain the track bed; a grassy way by a line of telegraph poles. To the right, below Hattonknowe, is a cluster of holiday huts.

A handsome metal bridge (similar to the one upstream) spans the Water, after which the field edge track bed is little more than a raised way between Water and road. Eventually a path leads to a flood protection embankment which, in curving away from the meandering and flood prone Water, goes close to the A703. Beyond a gate, another path winds through a new plantation and leads to a fenced off area (warning, dogs run free). Turn right to regain the track bed. Continue to the overbridge, spanning both track bed and Water, that carries the road to Cringletie House Hotel. Latterly, tracing the line of the track bed, keep to the right of the few houses to reach the road from Redscarhead. (In the village is a memorial to George Meikle Kemp who designed the Scott Monument in Edinburgh.)

Head left for a few yards to where the line crossed the road (no trace now of the level crossing) and then spanned the Water. Only the bridge abutments remain; but all is not lost. Follow the road to the right, cross or ford the Water, whence a farm track leads to the track

bed on a modest embankment. At a wooden gate, follow the signed path to the left; a charming detour by the banks of the Water, which avoids the curtilage of Kidston Mill. Rejoin the track bed and cross the Water yet again—the bridge deck has long planks--to reach a minor road.

Follow the field edge to the left and then rejoin the track bed, slightly overgrown, which slants towards the Water and another missing bridge; albeit the abutments remain. Splashing over bare-foot is possible, but upstream is a farm ford where, in dry seasons, boots-off may not be required.

The far side modest embankment leads to a ditch and a burn, detoured to the right. To the left is a caravan park and to the right new houses, the outskirts of Peebles. The line of the railway is lost, but a well-used local walk leads through a slightly wooded area. The path turns right by a wooden bridge over a burn, then left and left again by a metal footbridge for the final crossing of the Eddleston Water, known locally as the Cuddy and home to many a duck.

Turn right on the tarmac path, signed Peebles town centre via The Cuddy. Turn left by Cuddyside and pass through the Sainsbury supermarket, if open, to reach a large car park in the triangular area between the A703 and the Eastgate, the site of Peebles East station and goods yard.

Only the goods-cum-weighbridge office remains. The yard was curiously laid out, with several sidings cut by the little-used connecting line from Peebles West.

Even the disappointing finale will not spoil this interesting 10-mile stroll. Generally accepted to come from the Brythonic *pebyll* (tents), signifying a temporary settlement, Peebles is a popular tourist destination, home to many who commute to Edinburgh and a worthy place to rest before continuing to Galashiels.

Carstairs-Dolphinton-Leadburn

The Caledonian Carstairs—Dolphinton line (with intermediate stations at Bankhead, Newbigging and Dunsyre) and the North British Dolphinton—Leadburn line (with intermediate stations at Broomlee (for West Linton), Macbie Hill and Lamancha) never provided a through passenger service. Attempts to have a common station at Dolphinton, shared by both companies, were unsuccessful, so bizarrely the small village ended up with two stations only some 500 yards apart.

Nevertheless, apart from an abundance of electric fences, particularly on the Bankhead/ Newbiggingmill section, the combined track beds offer a pleasing 22½ mile walk.

Maps 72 and 73 are required.

Carstairs Junction to Bankhead station 1½ miles

East of Carstairs Junction is a triangle of extant railway lines. From the NE side, the start of the Dolphinton line is briefly mapped then lost on approaching Lampits.

From there can be seen a mapped, gracefully curving, embankment which, in bypassing Carstairs station, gave a through line from the South to Edinburgh. Known as the Queen`s Curve, this was allegedly built because Queen Victoria had made known her displeasure in having to enter the "black hole" of the original station at Carstairs. The more prosaic explanation for its abandonment (1860) is that topping-and-tailing of passenger trains at Carstairs, necessitating reversals, was the general rule from the opening of the Caledonian Railway. Moreover most goods trains shunted at Carstairs. In short, the curve was little used. The installation of today`s much tighter Strawfrank curve (1872), which re-introduced direct east-to-south working, was probably the result of expanding traffic after the amalgamations of the mid-1860s.

From map ref 964455, follow the Lampits tarmac road, pass by the abutments of the railway bridge, and turn left to reach a farm track in a grassy cutting that gives delightful walking. Cross the Carnwath Burn and pass close by the upper Clyde. (Further south, the river gives the impression of a large moat round Tinto Hill.)

All good things come to an end. Approach another cutting whose far end is wet and blocked off. However, follow the field edge on the north side to reach the high abutments

Dolphinton; link line passing behind NBR station *Lens*

of the demolished bridge over the B7016. Bankhead station house, on the far side, is now a private residence and other new houses occupy the site of the station of which there are no other remains.

Bankhead to Newbigging station 3 miles

Given the numerous electrified wires to be crossed, a Montgomery fence straddler is all but essential on this section.

The comments of one companion on the day are apposite. "I enjoyed the walk, but that was due mainly to the company, the lovely weather and parts of the surrounding countryside. Frankly the once walkable line is now more akin to an obstacle course. It could not be recommended as a pleasant country walk. Only a railway historian or explorer, anxious to find out more about the route, might feel motivated to complete the course."

However, if still motivated and carrying a fence straddler, head south on the B7016 beyond the bridge, then follow the field edge to the left to regain the track bed. This is cattle country so take heed of bulls and cows with calves.

Indicative of the terrain to be crossed, there are many attractive stone-arched bridges; on to Dunsyre the majority are still extant. The Caledonian Railway often built lavishly—the bridge across the Westruther Burn being a fine example. The few missing bridges are easily bypassed.

The sequence of modest cuttings and embankments continues into sheep country. Despite many a gate to cross (beware the electrified wires) the grazed track bed gives easy walking. Pass a clay pigeon shooting site to the left, then to the right is the meandering Medwin Water. Later comes a secluded way where the grassy track bed cuts through a plantation. Beyond another fine arched bridge is a cutting where a pond has formed, but that is easily side-stepped, and so to the site of Newbiggingmill station just west of the A721.

A loading bank on the left by sheep pens can be identified but no other traces of the station remain. Curve left on the former station entrance to reach the road, where only the grassy embankments of the underbridge remain.

Newbigging to Dunsyre station 4½ miles

Regain the far side embankment. Beyond another fine arched bridge, the track bed is used as a silage store, at which point follow the right hand rim. To the south is an extensive area of gravel pit workings which later extend to the side of the track bed with a sign warning of the quarry workings.

Involving more fence and gate crossings, the long curving stretch of track bed hugs marginally higher ground on the northern side of the bleak, boggy moorland. The lack of mapped contours indicates the flatness of the area where the South Medwin Water flows in an almost straight drainage ditch. (The Clyde/Tweed watershed is curiously hard to determine). However, the oddly beautiful moorland traverse gives a grand setting from where to admire 516m Black Mount.

Although one dip might be wet, the slight embankment keeps the walker well above the boggy ground. A footbridge over a burn eases the way and, where a minor bridge has gone, the small stream presents no problems. Avoid a cutting filled with silage by keeping to the right by the field edge, and then regain the track bed.

The bridge over the road at Dunsyre is no more, but a sheep track on the right hand side of the embankment leads to the road. Turn left, then immediately right, as indicated by the sign to Dunsyre Mains. Here begins the signposted 20-mile Pentland Way from Dunsyre to Swanston, Edinburgh. This point is also signed, public footpath to Garvald 1¾ miles and West Linton 5½ miles.

On the right hand side is The Halt, the old stationmaster`s house, now much altered and extended as a private dwelling. I was privileged to have the owner show me old photographs and some rivets as mementos. No trace remains of the two small buildings on the platform, now encompassed within the garden, though the line of the vanished railway, parallel to the lane to Dunsyre House, is self evident,

Dunsyre to Dolphinton (Caledonian) station and Dolphinton (North British) station 2½ miles

Beyond Dunsyre the track bed is partially lost within a narrow plantation. A metal gate on the north side of the station gives access to the field edge by the trees. In due course turn right by another gate to regain the track bed where it emerges from a cutting. Thereafter, modest grassy embankments and shallow cuttings give a most pleasant walk through open country and the few fences to be crossed pose no problem.

Pass under a stone-arched bridge, map ref 085486, and continue to the bridge over the Medwin Water. Oh dear! The abutments remain but alas not the central span. No problem though. At times of low flow the narrow Water is easily forded (boots on) a short distance downstream. Even at times of higher flow the few yards can be crossed bare-foot; a tonic for weary feet.

(For those unaccountably unwilling to get their ankles wet, return to map ref 085486, turn north and follow the signed Pentland Way. When the track reaches a ford across the Medwin, follow the path on the left to cross the West Water by a footbridge. Later slant right by grassy path back to the Medwin and another footbridge, to reach Garvald. Continue to the road bridge over the track bed at map ref 102483.)

The next minor bridge (over a farm track) has a missing central span but is easily detoured to the right, where a metal gate has a green disc with a white arrow sign. The track bed, gently curving towards Dolphinton, gives a lovely stroll through a wooded area. At a too-overgrown cutting, follow the farm track on the right to gain the road T-junction, map ref 102483 (and wait for those companions who did not ford the Medwin).

Follow the road SE for a few yards, and then turn left on an easy-to-miss narrow path to regain the track bed; very much a local walk, parallel to the road. Hidden from the walker`s view are two road signs denoting the South Lanarkshire/Scottish Borders boundary. The 200 yards distance between signs is apparently no man's land.

Continue as far as Karecole Farm Stables then revert to the road and pass the sign for Dolphinton. Sadly all evidence has gone of Dolphinton (Caledonian) station and coal yard; the site is now occupied by houses.

Pretend to be a train passenger obliged to walk to the rival company`s station, but in the knowledge that station-to-station transfers were actively discouraged and that, given the poor connections, a lengthy wait was likely! Head a short distance northwards by the A702 road side pavement. Carefully cross to the far side---the overbridge has been lost to road improvements. A lane, once the track bed, leads past one cottage to the still readily recognisable site of Dolphinton (North British) station. Respect the station house curtilage by going only a few yards down the lane.

On the right is the old engine shed and straight ahead is the station house, now signed Sunnybank, with the mandatory clock, this one stuck at 9.40am.The platform is on the south side of the station house and the track bed to the engine shed marked the terminus of the North British line, opened in 1864.

The Caledonian-to-North British connecting line is on the north side of the station house. It would appear that this connection was put in by the Caledonian before their branch was inspected by the Board of Trade, prior to opening in 1867. The inspector stipulated that it was to be used only for goods and mineral traffic. Horse-boxes (technically coaching stock) are the only "passenger" vehicles known (from the record) to have been transferred company-to-company.

Modern houses occupy the adjacent site of the Old Creamery, from where milk was transported by train to Glasgow. With reduced demand for milk during Glasgow Fair week, the surplus was used to make cheese. Under the LNER, when the LL&D closed to

passengers, milk and livestock traffic from Broomlee and other intermediate stations was taken via Carstairs to Edinburgh, at least for a time, by arrangement with the LMS.

Dolphinton to Broomlee (for West Linton) station 4½ miles

The regular bus service between Dolphinton and West Linton may ease transport arrangements.

To avoid the station house and the lovely garden, walkers continuing eastward should follow the A702 and then regain the track bed east of the private site.

At a gate, turn right on the mapped path by the strip of trees, and then turn left by another gate. Beyond the wide gap between the trees, a line of fence posts identifies the scarcely discernible grassy track bed. Later, the track bed, now on a slight embankment, crosses a burn. Take care on the wooden span.

The long straight grassy way crosses the lane to Felton Cottage. Beyond, unmapped trees cover the track bed, but a winding path permits progress; and so to Medwyn Mains, overlooked by 451m Mendick Hill. The latter name may refer to some long forgotten skirmish between the then inhabitants and Roman troops. (The course of a Roman road runs between the hill and the A702.)

The track bed, briefly a broad farm lane, crosses the farm road to Hyndfordwell, then continues in grassier form, curving round the south side of White Moss. Pass by a gate with a sign, warning of shooting in progress. Stroll past an old garden roller at the entrance to a house and then the abutments of a vanished accommodation bridge. A tarmac drive by new houses leads to the track to Whitemoss Farm, whence the track bed, parallel to the minor road from Romannobridge, becomes indistinct. It is easier to follow the road north to Bogs Bank and the entrance to an old estate drive, map ref 152503.

(Further north, the railway bridge over the road has gone, though part of the west-side approach embankment can be seen. The indistinct track bed goes between a sewage treatment works and the local cemetery and by paddocks, onwards to the Lyne Water, where only the abutments of the former girder bridge remain. Following the river downstream is definitely not recommended.)

The drive leads to a path which curves right by the mandatory rhododendrons, then left to a beautiful estate bridge. Cross and immediately turn left; follow the woodland path by the east bank of the Lyne Water and then the Cairn Burn, thus skirting Broomlee Outdoor Education Centre. This is private land, so keep dogs on lead. Reach a lodge, turn left at the stone pillars and follow the B7059 pavement to the West Linton road sign and the handful of houses. Here stands the much altered Broomlee station, which can also be reached from the village on Station Road, otherwise the B7059, by the east-side pavement.

By a curious coincidence my collaborator claims a close connection with Broomlee

station, the teenage home of his wife Christine. Her father began the conversion of the combined station house and stationmaster's cottage into today's modern dwelling – which subsequent owners have continued, much reducing the railway character of the building. John recalls how in the 1960s he habitually asked for the bus stop "at the station" – though passenger traffic had ceased thirty years earlier. That would not be an intelligible request today.

A small garden covers the track bed. All trace of level crossing and signal box (prominent in old photographs) has gone. However, on the far side is a building, tastefully converted into a private residence, signed West Linton The Old Engine Shed est 1862; actually a goods shed. The line was generally worked by a single locomotive stabled at Dolphinton.

West Linton, the settlement by the Lyne Water, in the pre-1914 years became a frequent destination for excursions, Sunday-school picnics and the like. And there was once a "commuter" through coach to and from Edinburgh, shunted at Leadburn.

Broomlee to Macbie Hill station 2 miles

Beyond the coal yard on the east side of the B7059, the track bed embankment leads to a missing span over the awkward Cairn Burn. Cue for a detour. Head south past the lodge and then turn left (east) to a small housing development. A track on the right, skirting the houses, leads to two unfriendly high gates which must be crossed to regain the track bed.

On a lovely curving and gently ascending route, below the radio mast on Broomlee Hill, the track bed eases away from the Cairn Burn. Three successive planked spans and a few barbed wire fences demand care. Now, with lines of telegraph poles on both sides, approach a prominent stone arched accommodation bridge with a discarded warning sign, laden weight not to exceed 13 tons.

On the right is an unmapped stone pillar some 12 foot high, which, given the rising ground to the east, appears to have served as an unconventional trig point. Beyond the bridge, traverse moorland and pass by the south side of a plantation to reach a modest cutting. Macbiehill`s stationmaster`s house comes into view on the right.

No other station buildings remain. All is grassy and overgrown save for a remnant of platform edge on the one hand and traces of a brick wall, overlooking a tributary of the Cairn Burn, on the other. A short mossy track on the right curves up to the Whitfield—Macbiehill road. Macbie Hill (sic) Station House, nowadays a charming private dwelling, lies beyond a blue gate. There are only a handful of houses in the immediate area and the hamlet of Macbiehill lies one mile to the south.

Macbie Hill to Lamancha station 1½ miles

Return to the track bed.

Although the line closed to passengers and freight in 1933, the Broomlee—Leadburn section was reopened during World War II, as part of the Royal Navy Armaments Leadburn Depot, to transport munitions to well-separated storage sheds alongside the track; a lengthy stretch from just west of Leadburn Junction and on to Macbie Hill station. Little evidence now remains. As happened elsewhere, the line was also used for storing redundant locomotives. It was not until 1960/1961 that the track was finally lifted.

With Deepsyke Forest to the left and moorland to the right, it is an easy and pretty stroll beyond the road bridge. At the SE end of the forest is a sign, "Caution, this is a working peat bog; please be aware of machinery movement, deep drainage ditches and soft ground". This refers to the ground to the left.

After a roughish area, the track bed straight ahead becomes little more than a grassy way by the fenced edge of pasture land (this is predominately sheep country) to reach a high embankment where one small gap is easily detoured. The track bed then gradually curves to yet another stone-arched accommodation bridge, by a solitary tree. Then a lovely mown way gives a grand entrance to Lamancha station.

An outdoor riding school and a large barn occupy the station yard. The original station building, so much altered and enlarged that its railway character has gone, dominates the old platform on the right. Continue to the A701 (note the station sign) and turn right a short distance to the entrance to the stationmaster's house, tucked away and difficult to photograph. A sign in the porch proclaims Station House 1865. The hamlet of Lamancha, straddling the A701, lies one mile to the south. The derivation of the name is obscure.

Lamancha to Leadburn station 3 miles

From the entrance to the stationmaster`s house, cross the A701 with care, head towards the obvious arched bridge and follow the line of the track bed. Slant right to the slight grassy embankment, from where it is an easy stroll to the abutments of the vanished overbridge on the tarmac road linking the A701 and the A703.

An easy detour regains the track bed which rises gently towards Leadburn. The tributaries of the Lyne Water, flowing to the Tweed, have been left behind. Ahead, parallel to the A701, is the Lead Burn (the outflow from Whim Pond), which, via the Black Burn and the valley of the North Esk, flows ultimately to the Firth of Forth.

Pass by a modest cutting and detour a poultry farm. The few barbed wire fences to be crossed scarcely require a fence straddler. The indeterminate track bed traverses a flat area, reed covered to the right, but then curves slightly right to pass through a narrow strip of wood. At the far end is a cross track; after rain, the intersection is liable to be churned up by cattle. However, a modest embankment then leads past the boggy ground, and to the right can be heard traffic on the busy but unseen A703.

The grassy track bed traverses attractive moorland, once the site of a Forestry Commission conifer plantation. The Commission chose to sell the site after the timber had been extracted. In 2007 the Friends of Leadburn Community Woodland purchased this 17 hectare site as a raised bog restoration area. This long-term project deserves support. (Those

Leadburn in NBR condition: the name board reads 'Change carriages for Lamancha, MacBiehill, Broomlee & Dolphinton' *Lens*

who do not undertake the walk from Lamancha will find it worthwhile to approach the Community Woodland from Leadburn.) The semi-open moorland gives a most beautiful stroll to the trailing junction with the Hardengreen-Peebles railway.

Finally, a winding path leads to the embankment overlooking the A703. Descend to the road and cross with care. Now for a brief visit to Midlothian. As described in the introduction to the Leadburn-Peebles Walk, a short lane on the left gives access to the far side embankment and the site of Leadburn station.

Symington-Biggar-Peebles

With three sizeable viaducts and a lengthy tunnel, the 19-mile line, latterly following the Tweed, has much to offer the walker.

At the time of writing, the Upper Tweed Railway Paths group has plans to convert much of the line into a waymarked route for walkers, cyclists and horse riders (cf. the already completed track bed path from Peebles to Innerleithen). At least for walkers, a waymarked way is hardly necessary. The route is perfectly walkable, detours are minor and any "sanitisation" would spoil that feeling of getting away from it all.

Maps 72 and 73 are required.

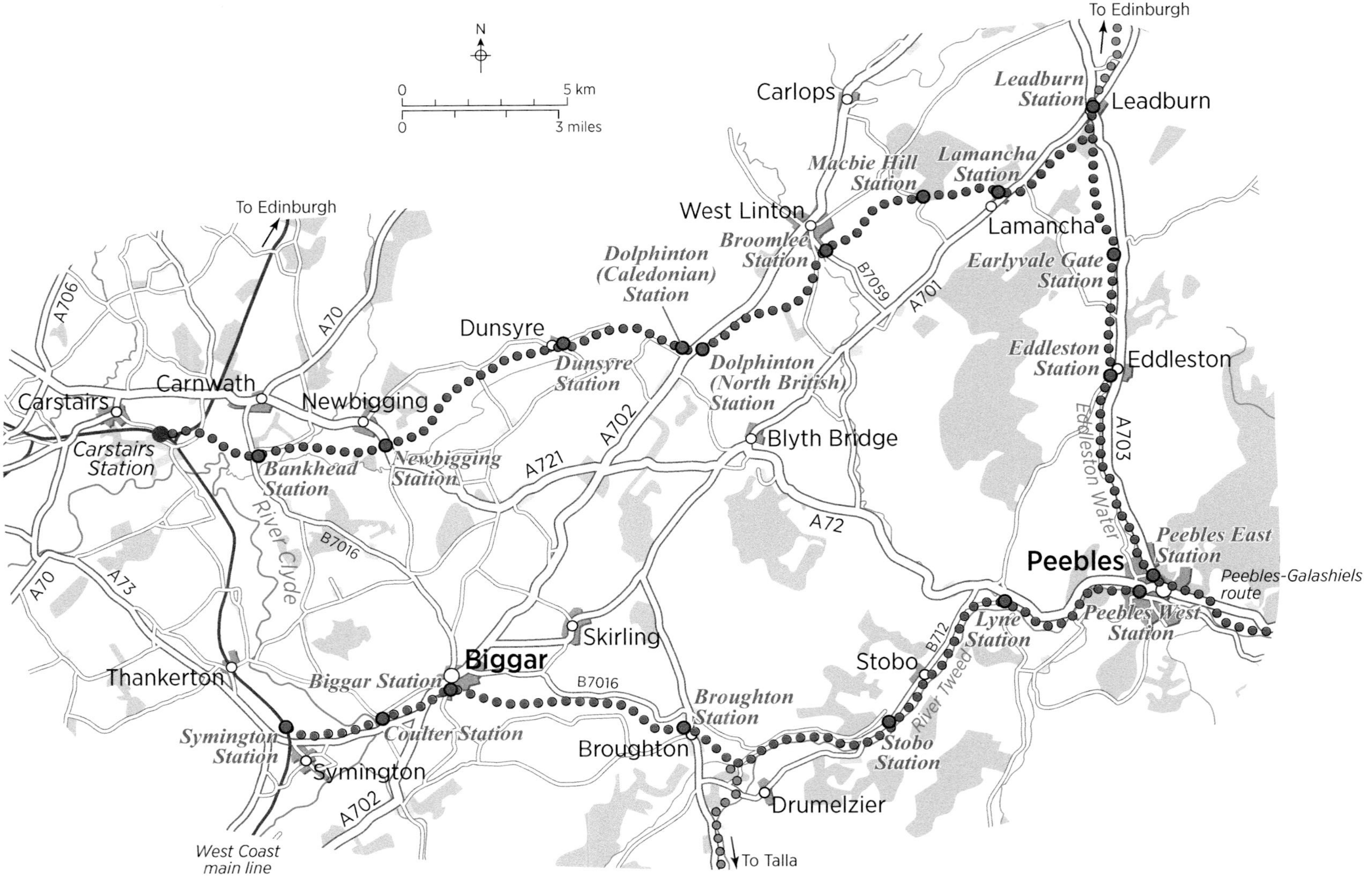

N
0
5 km
0
3 miles
To Edinburgh
Carlops
Leadburn Station
Leadburn
Macbie Hill Station
Lamancha Station
West Linton
Lamancha
Broomlee Station
B7059
A701
Earlyvale Gate Station
Dolphinton (Caledonian) Station
A706
A70
Dunsyre
Dunsyre Station
Dolphinton (North British) Station
Eddleston Station
Eddleston
Carstairs
Carnwath
Newbigging
Carstairs Station
Bankhead Station
Newbigging Station
A702
A721
Blyth Bridge
Eddleston Water
A703
River Clyde
B7016
A72
Peebles
Peebles East Station
Peebles-Galashiels route
A70
A73
Lyne Station
Peebles West Station
Skirling
Stobo
B712
Thankerton
Biggar
B7016
Biggar Station
Broughton Station
River Tweed
Symington Station
Coulter Station
Broughton
Stobo Station
Symington
A702
Drumelzier
West Coast main line
To Talla

Piers of Coulter viaduct *Furnevel*

Symington to Coulter station 2 miles

Symington lies between the eastern slopes of Tinto Hill and the meandering upper Clyde. Pronounced *Sigh*mington, the village was named after Simon Loccard, one of a number of sheep farmers from Flanders who were given land in Clydesdale by the King of Scots c.1155. In railway terms Symington was important not for its own sake but as a junction and, like similar places, lost its station when the branch line closed.

The A72 passes under the Glasgow-Carlisle main line at the north end of Symington. On the west side of the bridge, a lane leads north to the junction site (unrecognisable as such). On the east side of the bridge, another lane leads to a derelict industrial area whence the line curved east. Today the track bed is soon lost to gardens and houses.

Head east on the A72 to the far end of the houses. Turn left at crossroads on a track (this heads towards Annieston) and immediately turn right. The track bed runs straight ahead, a lovely grassy way. Successive small stiles enhance progress with enjoyment not unduly marred by the close-by and busy A72. The few roughish sections are short, until a small but overgrown cutting. This may force the walker to make a temporary diversion to the narrow road side verge (take considerable care!). Regain the track bed and go under the Eastfield access road, from where a grassy line leads to the approach embankment of the lengthy viaduct across the Clyde. All the spans have gone, leaving a sad, forlorn line of six piers between the abutment and the river. Only one pier, with a small tree atop, stands in the water.

A handy path with stiles, used by fishermen, by the impressive multi-arched road bridge over the Clyde, leads to the A72. Cross the bridge and turn left on Cormiston Road by the

signed Historic Scotland Coulter Motte Hill. Though lacking a fortified timber tower on the summit, Coulter Motte is an impressive grassy mound. A superb viewpoint, do climb to the top.

Hard up against the Motte is the Coulter station master`s house. Now a general storage yard, the station site still retains a long platform edge. No evidence remains of the level-crossing. I was privileged to talk to an elderly gentleman who turned out to be the son-in-law of the last stationmaster.

The village of Coulter (also Culter, though the pronunciation is "Cooter") lies 1½ miles south of the station where the A702 crosses the Culter Water.

Coulter to Biggar station 1½ miles

Return to the A72, and from the east side of Wolfclyde farm, turn left on a broad way, used for tractor access, to regain the track bed. It is then a short, but enjoyable, easy stroll between fields where I was told to look out for some peculiar white pitted ballast. This turned out to be remnants of furnace slag with numerous gaseous holes. Later, I discovered that such ballast was indeed quite common.

A grassier way on a modest embankment leads to the filled-in overbridge just north of the A72/A702 junction. On the far side, beyond the remains of a little overbridge spanning a ditch, the track bed slanted NE; an area now subsumed by the southern suburbs of Biggar.

Follow the roadside pavement into Biggar and turn right on Station Road. A few yards beyond Boghall Avenue, at a slight rise in the road, the overbridge has been filled in and new houses cover the line on both sides. To the east can be seen the top of the station signal box, stranded amid an industrial estate.

Passing brick-built railway cottages, walk back up Station Road and enter the industrial estate where the ruinous station building by the signal box is a sad sight. Though the railway

has gone, the town`s industrial past is recalled at the Biggar Gasworks Museum; one of the first small-town coal gasworks to open in Scotland (1839) and among the last to close (1973).

Biggar to Broughton station 4½ miles

With no way through beyond the far end of the estate, continue up Station Road. Before reaching the main road, turn right on Market Road for a few yards. Note on the right a marker post for the charming path by the side of the Biggar Burn.

Pass Biggar High School and the western end of Biggar Golf Course, then cross the Burn by a wooden bridge to gain the track bed embankment heading east. (To the west it is gated off and overgrown.)

The track bed re-crosses the Burn by a concrete bridge at which is a sign pointing back the way to Biggar via Station Road, part of the Biggar Country Paths Network. On the left are the golf course car park and an attractive pond, worthy of a detour, beside the caravan park. The slightly raised track bed, very much a local walk, continues past the golf course and, once more crossing the Burn, passes from South Lanarkshire to the Borders.

What follows is a superb stroll through part of the Biggar Gap, probably a prehistoric glacial breach that separates the Pentlands from the Culter Hills and gives easy passage from the Clyde to the Tweed. Easing the way through reed-covered terrain, the track bed runs on the south side of the Biggar Water; albeit at first more like a broad drainage ditch. Unfortunately, from the walkers` point of view, the straight stretch immediately ahead, also used by horse riders, may be churned up and muddy after rain, whereas full of ice-bound ruts in winter.

Rachan, looking back to Broughton, with Talla Light Railway trackbed to left ***Crawford***

Pass through a slight cutting after which the track bed curves right then left. The Biggar Water, fed by drainage ditches, including the Kilbucho Burn, is now full flowing on its gentle descent towards Broughton. I was lucky on my January day to see lots of geese.

Pass by the Scottish Water sewage works to a large metal gate; the entrance to the site of Broughton station, now a general purpose yard. On the left side of the gate is a sign, Biggar 8km. Eastwards, another sign reads, Broughton, John Buchan Way. Named after the author-statesman who had many associations with the area, the 13-mile, well-signposted Way runs from the John Buchan Centre in Broughton to Peebles.

To detour the yard, cross the Water by footbridge and skirt the perimeter of Broughton Brewery to reach the A701 where a sign proclaims, "Beer sold here, shop hours Monday to Friday 10am to 5pm."

Turn right at the entrance to the yard. (There is a guard dog when no workers are present.) No station buildings remain though one lengthy platform edge is readily identified. The far end of a very wet cutting leads to the overbridge on the A701, beyond which is a paddock within the curtilage of a large house.

It was my good fortune to be given both a guided tour round the station remnants and a history of the station`s connection with Talla Reservoir.

Talla Reservoir was built to cope with Edinburgh`s burgeoning demand for water. Existing roads were unsuitable to move the considerable volume of materials, such as more than 100,000 tons of puddle clay (a mixture of clay, gravel and sand) needed to create a watertight seal. Accordingly, accessing the remote site necessitated the construction of a light railway. This shared the Broughton-Peebles formation to Rachan, then turned south along the Tweed and so to Victoria Lodge by the edge of the reservoir. (This arrangement, two single lines on a double track formation, was cheaper than building an outlying junction.

The Highland Railway/Great North of Scotland Railway example at Boat of Garten was well-known.)

Work started in 1895 and was completed in 1905, with Broughton station enlarged to cope with the traffic. The line also transported workmen and a platform was built at Crook Inn. The Talla line met the requirements of the 1896 Light Railways Act, but efforts to retain a public service did not succeed and the tracks were lifted in 1912. Water passes by gravity through a 35-mile network of pipes and tunnels and crosses seven valleys to reach treatment works at Alnwickhill and Fairmilehead on the south side of the city.

Broughton to Stobo station 4 miles

Whilst acknowledging the influence of my crisp, beautiful and windless snow-covered January day, I might well rank the Broughton-to-Stobo section, by the Biggar Water and the banks of the Tweed, as my most enjoyable station-to-station walk to date.

Cross the A701 and head south to the overbridge (and the adjacent pedestrian walkway) from where the cutting is well seen. Return to the car park opposite Broughton Primary School and turn down the lane by the side of the Biggar Water; its flow enhanced by the Broughton Burn. Pass the bowling and tennis clubs, turn right by the Easter Calzeat sign, and cross the convenient footbridge over the Water to a winding path on the track bed. Re-cross the Water, from where a beautiful grassy embankment, used for farm access, gives superb walking.

Just before a cutting is the junction where the Talla Railway curved right on a broad grassy way to the approach embankment of the vanished railway bridge over the Water.

Only the abutments remain. Retrace steps for a few yards and follow a path that slants back to the track bed at the far end of the cutting, and so to the underbridge by Rachan Home Farm. The beautiful grassy embankment continues as a raised walkway bisecting boggy ground, and leads to a fisherman`s hut which overlooks where the Biggar Water joins the Tweed. The river is then followed all the way downstream to Peebles.

Initially pleasantly tree-fringed, the Tweed-side stretch leads to a rougher area, traversed by winding sheep tracks. (If snow-covered, beware a cattle creep with missing slats.)

Later the track bed, used for farm access, passes by a large pond on the right. Ahead is an overgrown cutting, blocked at the far end by a farm building. Slant right by the mapped track, cross the bridge over the cutting and pass the farm building. Turn right to regain the track bed, where silver birches grace a curving embankment.

Reach the B712 where the overbridge has been lost to road realignment. The old B712 (giving access to Crownhead Cottage) curves left over the cutting. Follow the B712 verge to just beyond the junction with the minor road to Dreva. A path slanting right leads back to the track bed, parallel to the road. Beyond a cottage is the site of Stobo station. Turn left to the road and pass by the Old Smithy. The second building opposite the post box is the dilapidated station house and on its far side is a surviving platform edge. To the east, the station yard, though recognisable, is occupied by gardens of the new houses that line the road.

The hamlet of Stobo, which straggles along the B712, has an unusual name; derived from the Old English pre-7th Century *stubb,* tree stump, and either *holh,* hole, hollow depression, or *halh,* nook, recess.

Stobo to Lyne station 3½ miles

Regain the track bed at the east end of the string of houses. On the right, a bridge across the Tweed gives access to Dawyck Mill. Cross the Weston Burn and on the left is Stobo Castle; a

19th century castellated mansion now transformed into a luxury hotel-cum health spa. The track bed, a lovely raised walkway clear of boggy ground on both sides, passes by the Easter Dawyck bridge and crosses the Easton Burn.

A rougher section leads to a burn and a winding path, at the end of which a short detour to the B712 is required. On regaining the track bed, note the sign, "Salmon fishing, strictly private.... this water does not belong to Dawyck or Tyneside Anglers". The Tweed is internationally famous for its salmon. Traverse a partly filled in area. Descend to a long cutting, which, beyond an accommodation overbridge, becomes shallow and grassy. Make a minor detour to the Tweed and thus avoid a decayed cattle creep.

Hugging the Tweed, the old railway gradually curves away from the B712. A barbed wire fence (straddler to hand!) must be crossed to enter a wooded area where the track bed shrinks to a winding path. Beyond another fence is a small dump, after which a grassy tree-lined cutting gives a superb walk close by the river.

Pass through a gate from where the tree-interspersed track bed area widens. Beyond another gate, buildings block the way to the site of Lyne station. Turn right, cross another fence (barbed wire again!), and follow the field edge to a gate and a minor road. Turn left to Lyne Station, so mapped; a pleasing settlement where new houses now predominate. (Mapped Balquhidder Station is another instance where a vanished railway or station is thus kept in memory (refer to Dunblane—Crianlarich Walk.)

Pass by the lovely unmistakeable station house, identical to, but in better order than, that at Stobo. One of the twin protruding frontages with bay windows has a tasteful glass extension. Though no other structures remain, the platform edge can be seen within a garden. Continue to the Lyne viaduct which also spans the minor road.

The Lyne Water, the only Pentlands stream draining to the Tweed, gives West Linton its rather fragile Borders identity, exemplified in the traditional festival called the Whipman Play, cognate with the distinctive annual celebrations held in every Borders town. Despite the West Linton connection, Lyne is of greater antiquity and is pronounced "Line" (refer to Carstairs—Leadburn Walk).

Steps on the left, signed, Tweed walk, Peebles Lyne Station, give access to the approach embankment of the 133 ft long skew viaduct. Its three well-proportioned masonry arches are a foretaste of the larger Neidpath viaduct downstream. The station entrance drive, north of the viaduct, is opposite a cottage.

Lyne to Peebles 3 miles

The final three miles to Peebles make another must-do walk.

Once over the Lyne viaduct, the lovely mossy track bed curves right and provides a semi-wooded walk between the A72 and the Tweed. It later becomes a well-worn way used for farm access.

Cross a small burn south of Edston and pass by the remnants of a surfacemen`s brick hut. At a fence which divides the track bed, keep to the left—the true railway alignment. Reach a missing underbridge—only the abutments remain—from where steps on the right lead down to the Manor Valley road, north of the old Manor Brig that spans the Tweed. A farm track leads back to the track bed.

With the steep slopes of South Park Wood to the right, approach a narrowing gorge, shared by road, river and former railway. Traffic on the constricted road is controlled by one-way lights.

A gently-rising approach embankment indicates that it is time to cross the Tweed.... and the first sight of the graceful Neidpath viaduct will put a spring in the step for even the most jaded walker. 440 yards in length, the A-listed curving structure consists of eight skew arches. Four substantial piers, designed with floods in mind, stand in the river which the railway spanned obliquely before curving into Neidpath tunnel. On the west bank, before crossing, is a triple signpost; straight on via the viaduct to Peebles, back to Lyne Station, and left to Neidpath Castle, as yet hidden by the curving gorge.

A short cutting leads to the B-listed Neidpath (or South Park) Tunnel. Necessitated by the encroaching steep hillside of South Park Wood, the tunnel, brick-lined for most of its 600 yards, takes a southerly curve some 150 yards from the far end, approximately conforming to the bend in the Tweed. It is a safe passage, albeit with a few drips, but a torch is well-nigh essential.

(On a previous visit, I found the tunnel entrance boarded up and took a rough path to the left, round the hillside. Though beset with tree roots, this offered impressive views of the Castle across the water. Of late 14th century origin, the building is not open to the public, except by arrangement, though weddings and other events are held there all year round.)

The tree-lined cutting at the exit to the tunnel may be muddy, but thereafter it is a riverside stroll into Peebles, past the graceful steel-truss Fotheringham pedestrian bridge--a gift from a former mayor of Johannesburg. Urbanisation has spread up-river, from the B7062 road bridge over the Tweed, to overwhelm the site of the "Caley station", ultimately styled Peebles West under British Railways. No station buildings survive. Keep to the right,

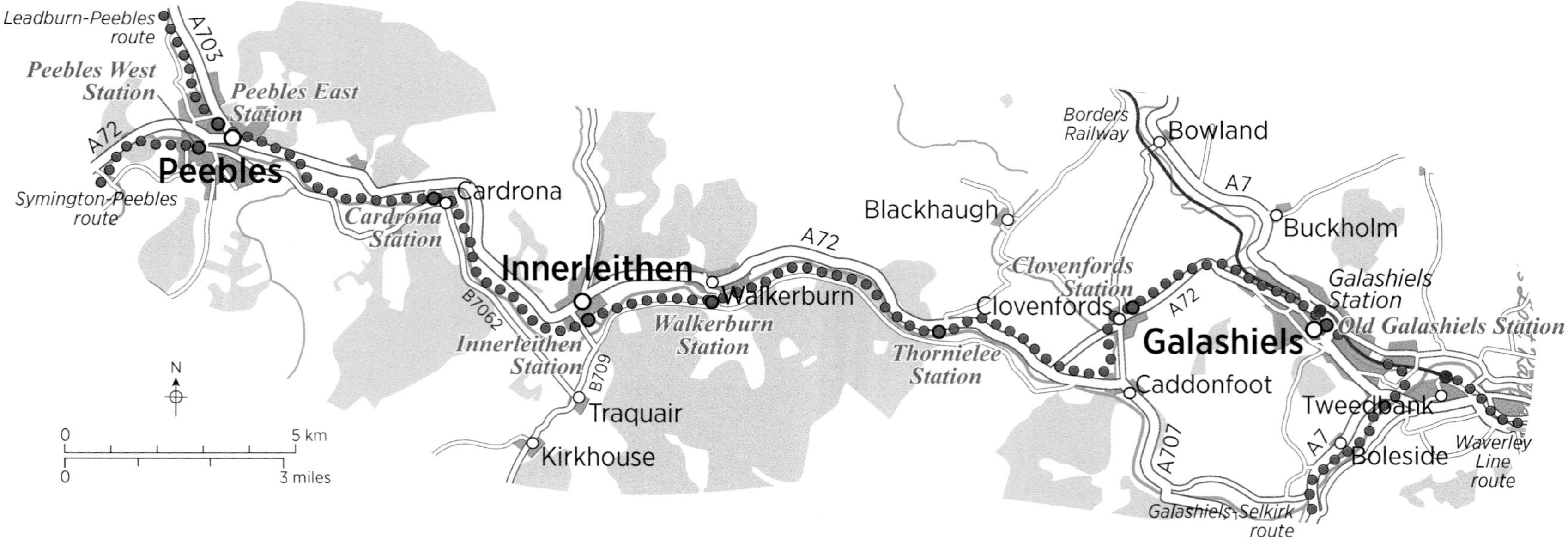
Leadburn-Peebles route
A703
Peebles West Station
Peebles East Station
A72
Peebles
Symington-Peebles route
Cardrona
Cardrona Station
Innerleithen
B7062
Innerleithen Station
B709
Traquair
Kirkhouse
Walkerburn
Walkerburn Station
A72
Blackhaugh
Thornielee Station
Clovenfords Station
Clovenfords
Borders Railway
Bowland
A7
Buckholm
Galashiels Station
Old Galashiels Station
Galashiels
Caddonfoot
A707
Tweedbank
A7
Boleside
Waverley Line route
Galashiels-Selkirk route
N
0
5 km
0
3 miles

away from the Tweed, and turn left on South Park West, by a line of brick railway cottages now white-painted, after which a lane leads to the riverside path.

Though possibly a sad ending, there is a post-script to the main walk from Symington.

Turn right (downstream) on the riverside path. Pass under the B7062 to find the last remains of the contentious "extension" to Peebles East station. Beyond a parking area, a grassy embankment, easily mistaken for flood protection works, continues to the Priorsford footbridge, just beyond which the junction line crossed the Tweed. Railway politics dictated that it was limited to freight transfer. The viaduct`s position and oblique alignment can still be determined—a portion of the far bank abutment endures—and the remains of the piers are detectable when the river is low.

Until the new diesel units arrived, Peebles offered a curious postscript to the long years of steam operation and inter-company rivalry. Though goods workings became vestigial, locomotives still used the junction spur to access the old Caledonian turntable after the former North British engine shed had been abandoned.

Peebles-Galashiels

The Edinburgh-Peebles-Galashiels route designated a modernisation show-piece (see p.6) but did not long survive. Nowadays, if travelling south by train, Galashiels station, on the Borders railway, is a good starting point from where to head west to Peebles, and the Peebles/ Galashiels regular bus service may assist with subsequent transport arrangements.

Map 73 is required.

Peebles to Cardrona station 3 miles

The track bed from the outskirts of Peebles to Cardrona (and on to Innerleithen) is now a tarmac path; a traffic-free route (apart from Cardrona village), ideal for wheelchair users, walkers, horse riders and cyclists. The path is part of the Sustrans National Cycle Network (NCN), signed routes which use a mix of off-road paths and quiet roads; it links to the Newcastle/Edinburgh section of NCN1 (the M1 of cycle routes) at Innerleithen.

From the large car park in the triangular area between the A703 and the Eastgate, cross the A703 and admire the impressive metallic sculpture of horse and rider jumping a fence; a tribute to Scott Brash, the world class show jumper from Peebles. The track bed, now a tarmac lane, terminates abruptly at an overbridge, beyond which are new houses and gardens.

Return to the A703 and from the roundabout head east by the A72 pavement to the main entrance of Peebles Hydro, one of two hydropathic hotels left in Scotland; the other is

at Crieff. Though new houses were being built at the time of writing, the vanished railway can be followed through the Hydro grounds to the east entrance and the abutments of the underbridge. (Alternatively, follow the A72 pavement to just short of the speed derestriction sign, and turn left.)

Follow the signposted Tweed Valley Railway Path which goes parallel to the A72 then, by way of a new concrete underpass, heads towards the Tweed. On the right are sewage works and on the left the town's once rail-served gasworks. Pass through Eshiels Community Woodland to reach a junction. Glentress, a forested area which has a series of world-class mountain-biking trails, lies ½ mile to the left, whereas straight ahead are Cardrona, 1mile, and Innerleithen, 3 miles.

The path continues through Cardrona golf course and crosses the Tweed by a bow-string girder bridge to reach Cardrona station. The construction of a modern road bridge over the Tweed led to the rapid development of the commuter village. However, Cardrona was originally a minor station on the south bank of the river when access from the Peebles—Galashiels road was by Horsbrugh Ford. Dating from the 16th century, Horsburgh—note the different spelling—is the ruined tower house well-seen from the A72.

The station site is easily identified by the platform and the wooden station building, waiting rooms and ticket office, now serving as Nashy's Coffee House. Custom has increased since the opening of the Peebles—Innerleithen path.

Later I met the owner of Innerleithen caravan park. His grandfather was the stationmaster at Cardrona and his grandmother the gatekeeper. Unfortunately the grandfather was killed in 1954, following an accident at the station.

Our meeting had a windfall—my informant owned an original copy of the plans and sections for the Peebles-Galashiels line, and this is now destined for the National Records of Scotland. Additions to the archive storehouse, not necessarily railway ones, often turn up casually, as my fortuitous encounter demonstrated. If any reader comes across similar material, please do consider directing it to the national archives.

Cardrona to Innerleithen station 2 miles

From the immediate roundabout, keep left along Cardrona Way and pass through Cardrona village. The name possibly comes from Cumbric—*caer* (fort) and *tronau* (of the thrones) or *trywnau* (on the hills). Look out for the sharp turn left, as signposted, to cross the track bed, then right on to the path leading to the Tweed. Little evidence remains of the railway viaduct, but help is at hand in the form of a footbridge a few yards upstream. Note that pedestrians have priority; horse riders please dismount!

Ignoring a junction on the left (Innerleithen by Caerleebank), follow the north bank all the way to Innerleithen. Enjoy the riverside stroll!

Before crossing the B709, Traquair Road, pass on the left what was a substantial goods yard, a private siding for the Waverley Mills. Thanks to the Leithen Water, the manufacture of high quality wool and yarns transformed Innerleithen village into a significant mill town. Sadly, with the decline of the industry in the late 20th century, mills began to close. Ballantyne's, the town's largest employer, was the last mill to do so—in 2010.

On the far side of the road is the stone-built station house, now a private residence; larger and more substantial than the modest Cardrona. The owner most kindly showed me around the site. (Perhaps indicative of tight Border communities, her sister is involved with the coffee house at Cardrona.) The house retains its attractive veranda and platform edge. The track bed is now a small strip of garden and further east an industrial estate blocks the way.

Innerleithen is known for its medicinal spring, made famous in Scott`s novel, St Ronan`s Well; mapped as St Ronan`s Wells. Innerleithen`s annual Borders Games include the Cleikum Ceremony in which St Ronan drives out the devil. Traquair House, south of the Tweed, and visited by twenty-seven kings, claims to be the oldest inhabited house in Scotland.

Innerleithen to Walkerburn station 2 miles

What follows is a complicated route through Innerleithen to regain the track bed. If unsure, keep asking for the footbridge across the Leithen Water or the Tweedside caravan park. (Alternatively, go along the A72 to the east end of the town and follow signs to the caravan park. A path on the west side leads to the track bed.)

Head down Traquair Road and turn left on Tweedbank Ley which leads to Tweedbank farm. Turn third left then right, as signed, Innerleithen Paths. The tarmac path turns sharp right and passes two staggered metal barriers to reach the footbridge. Cross the Leithen Water; turn right on Princes Street then left, to follow a line of telegraph poles on a tarmac path......the track bed!

Continue past the caravan park, from where a gentle rise leads to the six-arched lengthy viaduct obliquely spanning the Tweed; three of the piers are in the water. The signed

Walkerburn Path on the right goes under the viaduct and follows the north bank of the Tweed. However, cross the wooden-decked viaduct, after which the track bed, passing by paddocks, is a grassy way used for farm access.

The track bed deteriorates to a rough field-edge way, parallel to a minor road and some distance from the river; it then curves left across the large flat field. Where it is finally lost, continue on a minor road and pass an entrance to Plora Wood; this restored ancient woodland is a Site of Special Scientific Importance and the largest area of semi-natural oak wood in the district of Tweeddale.

Eventually turn sharp left on the road to Walkerburn. A lane on the left, before crossing the river, leads to a small industrial area; the extensive site of the former station yard where the principal traffic was coal for the Walkerburn mills. One stone-built store remains.

By the corner of the lane is the stationmaster`s house, now a private residence, Station Cottage. Close by, Old Railway Station is a white-painted building, substantially altered and extended across track bed and platform. On the wall is another mandatory clock, this one stuck at 9.25am. The full length of the platform edge can be seen from the station yard.

Taking its name from the Walker Burn, the small village of Walkerburn came into being in 1854 to house workers employed in Henry Ballantyne`s new tweed mill. The railway arrived 12 years later—but on the south bank of the Tweed.

Passengers were ferried across until a footbridge was built a year later. When floods made the Bold Ford impassable, horse-drawn carts and motor vehicles had to use the bridge at Innerleithen. Not until 1914, with the construction of the Caberston bridge, was direct access to the station achieved.

Closure of the Walkerburn mills has reduced the local population but the village remains an active community. A number of local businesses are sited in the old mill buildings and the village is home to the Scottish Museum of Textiles.

Walkerburn to Thornielee station 3½ miles

Were it not for the demolished railway viaduct downstream, the Walkerburn-Thornielee stretch would have been the line`s outstanding station-to-station walk.

No trace remains of the level crossing at Walkerburn. Traverse the grassy way adjacent to a rugby pitch and follow the track bed to the Bold Burn. The railway span has gone but a convenient foot bridge leads to the modest embankment on the far side, used by farm vehicles. To the left, along the Tweed, the effects of flood erosion are evident—and the old formation still safeguards the low-lying fields on the right. Pass a sign "Free range chickens, no free range dogs please." Giving delightful walking, the lovely curving embankment is soon hemmed in between steep wooded slopes and the high retaining wall that provides protection from the river.

At a horse-rider friendly gate, the track bed joins a gravely way from Elibank House. Reach a small wooden hut and a welcome bench by the river; a resting spot before searching for signs of the demolished railway viaduct. A few yards further on is a sign, Bridge Pool, and on the far bank can be seen a grassy embankment; but no trace remains of the viaduct, though the pier foundations can be identified when the water is low.

On my day, the owner of Elibank House was clearing overgrowth to give casting space for anglers; and an invitation ensued for coffee and biscuits. For generations the estate was the family seat of the Murrays. Elibank Castle, a 16th century ruined tower house, lies within the grounds and offers excellent views.

To gain the far bank, one option is to return to the bridge at Walkerburn— better, however, to follow the south-side road to Ashiesteel Bridge, sometimes called Low Peel Bridge, 3 miles downstream. Its graceful semi-elliptical arch, at 132ft one of the world`s largest rubble masonry spans, is a thing of beauty.

A pre-placed car may be the preferred mode of transport, heading west to reach the Thornielee Forestry Commission car park, whence the railway embankment can be regained via the A72 verge. (The curving road opposite the missing viaduct is hazardous for pedestrians. If dropped off, take great care.)

The track bed, exploited for farm access and power lines, gives a lovely grassy river-side stroll to the site of Thornielee (mapped as Thornylee) station. An odd location serving but a small community, it is yet another of the many railway outposts which were established with little expectation of worthwhile traffic, simply to ensure support or overcome opposition. Satisfying the local landowner was often the key reason. "The past is a foreign country, they do things differently there" – L.P.Hartley, The Go-Between (1953). What seem improbable locations today made sense in different social/economic conditions.

Beyond the small goods yard, the lengthy single platform leads to the combined stationmaster`s house and booking office; now a much extended private residence. Avoiding the curtilage of the house, a track on the left leads to the A72. A short distance further east, the station entrance goes past two railway cottages, one with a wooden Thornielee station bench.

Thornielee to Clovenfords station 3½ miles

At first strimmed, the track bed quickly becomes overgrown as it approaches the A72. Because of the road realignment at Thornielee, no evidence remains of the railway crossing point. Head beyond the slight dip in the road, and the power lines, to reach the eastern end of the hamlet. A short climb then regains the track bed; mapped but scarcely legible alongside the red line of the A72. Following the power lines—the path is vague-at-times—stroll above the noisy road towards a short cutting.

As if specially arranged for my day, workmen were strimming the track bed, so as to maintain access to the power lines; a welcome benefit for walkers. Nevertheless, the winding path is best tackled at times of low growth. Further on, the track bed, now terraced on the hillside, becomes overgrown. Although an indeterminate path encourages perseverance, descend at the first sight of the Nest roundabout, the A72/A707 junction.

(If not undertaking the traverse from Thornielee to the roundabout, travel by car or bus; walking on the busy A72 is not recommended.)

Road realignment has left no trace of the railway bridge over the A72, a few yards beyond the roundabout, but the high power lines record its alignment. Climb the far embankment

Clovenfords in LNER condition; vinery nearby (see Section *Thornieliee to Clovenfords) Lens*

and cross a dyke to regain the track bed and the power line poles, and where a faint path penetrates some stubborn briars. What follows is a most pleasant walk overlooking the road to Caddonfoot. A modest embankment gives good views of the Tweed, before the track bed curves left and crosses a farm lane by a telecommunications mast. Wave goodbye to the river; a constant presence since Peebles (or indeed from just east of Broughton).

Still curving left, a broad grassy way offers the first sight of the small village of Clovenfords. Regain the track bed at a brick-lined arched bridge by a line of old wooden fences. The next approach to the A72 has been lost to pasture. (If the field cannot be traversed, cut down on the right to a football pitch.)

Once again road realignment has obliterated all trace of the crossing point. On the far side, a tarmac lane (the old A72) crosses the Caddon Water and, as Millbank Road, winds its way to the roundabout at the A72/B710 junction. A few traces of abutments can be seen en route, but the track bed has been subsumed by houses and gardens.

Cross the roundabout to the Clovenfords Hotel, an 18th century inn on the then stagecoach route between Carlisle and Edinburgh. (Prior to its rapid 19th century growth, Galashiels was but a hamlet.) A prominent statue of Sir Walter Scott, who stayed at the inn in 1779, stands in the forecourt.

Continue on the A72, locally Vine Street; a reference to Tweed Vineyards, established in 1869 by William Thomson. Clovenfords was chosen because of its railway station (essential for delivering the many tons of coke required to heat the complex of hothouses) and because Thomson`s brother-in-law, a builder and contractor, lived in Galashiels.

From Vine Street, turn left on Station Yard, a new housing development. Easy to identify, the old station house has been extended but retains much of its railway character. At the rear, the garden covers the track bed. The houses at the end of Station Yard have subsumed the north end of the platform and the goods yard.

Clovenfords to Galashiels (Kilnknowe Junction) 3 miles

Return to the A72. On leaving the village, note the walkway sign, Galashiels railway station 3 miles. On the left a lane leads to part of Clovenfords goods yard but further progress is impossible. A house blocks the way and in any case the immediate track bed over the boggy saddle between Clovenfords and Galashiels is not walkable. Continue by the roadside pavement to a sign, 2½ miles to station. No trace remains of where the railway crossed the road and on the far side the unmapped track bed has largely gone.

Continue on the roadside pavement to a bend on the road (the discernible point where the track bed recrossed). From there a signed walkway directs the pedestrian to a tarmac path, in effect the final stretch of track bed. Much used locally, this charming path between the A72 and the Gala Water gives a lovely approach to Kilnknowe Junction. Pass by a sign, railway station 1½ miles. On the left can be seen the Borders Railway where it spans the Water. By now parallel to the new line, and as if heading towards the Eildons, approach a double-arched stone bridge. Just beyond is the site of Kilnknowe Junction, where the Hardengreen-Peebles-Galashiels loop rejoined the Waverley Route. No evidence remains of the signal box.

Continue to Wheatlands Road, spanned by the new rail bridge and follow the signs for the less attractive final ¾ mile to the new Galashiels station.

Fountainhall-Lauder

Between Fountainhall and Lauder there was one intermediate station, at the small village of Oxton. The line replaced a horse bus service (privately operated but subsidised by the North British Railway) linking Lauder with Stow station on the Waverley Route.

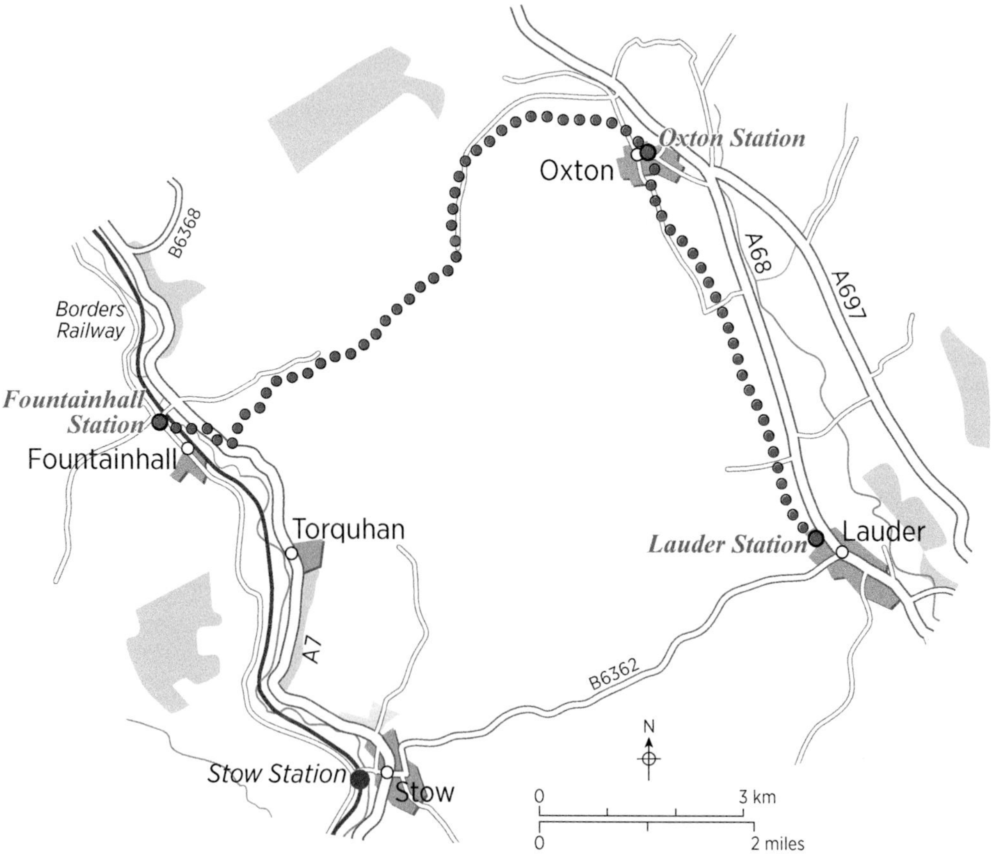

Fountainhall Junction: beyond the platforms and signals, the Lauder Light Railway diverged left *Lynn*

The line was steeply graded in places. To reach the 944ft summit, the watershed of the Nethertown and Mean Burns near Eastertoun, demanded a 2½ mile ascent from Fountainhall on a ruling gradient of 1 in 50. South from Oxton, the railway gently descended through Lauderdale, the valley of the Leader Water and today`s A68 trunk road. Lauderdale was on the Roman Road from York and remained for centuries the prime route between Edinburgh and England; it was followed by the armies of Edward 1 in the Wars of Independence and more generally by travellers and pilgrims, for whom the nearby Soutra Hospital provided shelter.

In a never heavily populated area, passenger use was always slender, albeit fishermen gained access to the many trout streams for which the area is famous. I met a gentleman whose grandfather had travelled on both the first train and the last. Given that several late-in-the-day conventional branch lines succumbed more quickly to road competition, it is a fair achievement that a passenger service lasted some 30 years and a goods service more than 50.

Simple sidings were provided for livestock traffic, principally sheep, to the marts at St Boswells and Gorgie, (Edinburgh). In World War Two, hardcore for airfield runways was sent by rail from Airhouse Quarry, west of Oxton. The Lauder line, which might well have succumbed to the 1948 floods, was granted a further decade of life because it served an Emergency Food Depot, established during the War and kept in being by the Ministry of Defence, in the Cold War atmosphere of that time.

To provide locomotives of sufficient power but limited weight was a challenge. (In British Railways days the line was worked by a tank engine coupled to a spare tender which spread the axle-load.)

Map 73 is required.

Summit level south of Oxton, under snow *Crawford*

Fountainhall Junction to Oxton station 4 miles

The hillside traverse to Oxton, completed all too soon, is a little gem. On my day I met one couple who had taken the bus to Oxton and were walking through to the A7 and the Edinburgh/Galashiels bus service. What a good idea!

Heading south on the A7, go beyond the signposted road to Fountainhall and Allanshaugh; road improvements in association with the Borders Railway have slightly changed the layout. Turn right on the old Allanshaugh road, and cross the Gala Water by the old bridge. Fountainhall station is boarded up, and the footbridge and platforms of this former junction have all gone—a forlorn contrast with the restored station at Stow.

Walk down the road, parallel to what was once the Waverley Route. It is good to hear and see the passing trains. Continue to the Still Burn and turn left across the field to the embankment that leads to the Allanshaugh railway bridge spanning the Gala Water. Though overgrown, the brick-lined skew-arched bridge gives easy passage to the A7. No trace remains of the level crossing. From the far side parking area, a wooden gate gives access to the grassy track bed parallel to the road.

Continue to a slight cutting, then a barbed wire fence and a rougher section. (This latter point can also be reached by walking further south by the A7 grass verge, then striking uphill.) Another short grassy cutting, curving away from the road, leads to a well-grazed embankment; this is very much sheep country. Also used for farm access, the sinuous track bed steadily rises. The modest cuttings and unassuming embankments exemplify "light"

construction (and such lines were often unfenced). Cross the track to Middletoun farm and on to the broad farm way (mapped) which climbs to the watershed, where the brick chimney of a surfacemen`s hut has survived.

Now for the steady descent by the infant Mean Burn; the mapped track bed now more akin to a grassy lane. Reach another railway hut beside which is a sign, back the way, path to Middletoun 3 km. Close by a ruin, cross a track (which climbs to Threeburnford and the start of a tarmac road) and head straight on. The track bed descends and then crosses the road. If the far side cattle-feeding area is muddy, briefly follow the road, and then regain the track bed; a charming embankment later cut into the hillside. The purist will not be tempted by the parallel road. Pass the disused Airhouse quarry and reach a metal gate, after which, close to the Hartside road junction, the track bed is lost.

It resumes beyond a wooden gate and across a side burn; the straight way, parallel to the Mean Burn, very much a local walk. Later, the burn curves sharply under the track bed. The unmapped railway bridge has seen better days; its brick side-walls are crumbling and the planked central span needs to be crossed with care. Later, the high fence surrounding a gas installation (Scotland Gas Networks Soutra Offtake) enforces a detour. Further on, reach an embankment; rather like a raised walkway between steep wooded slopes on the right and boggy ground on the left. Ahead is the first sight of the busy A68. Beyond a minor road, the track bed, curving SE, is cut into the hillside high above the Mean Burn and its confluence with the Headshaw Burn.

Reach a high fence, beyond which is debris from the site of Oxton station, cleared for a housing development (long planned but not yet begun). Detour right by the pasture edge, cross a small burn and follow the lower edge of the track bed to Station Road. The white-painted stationmaster`s house, a private residence at the corner of the road junction, is the only surviving structure. Beside the house is a post-Grouping sign, London & North Eastern Railway Warning to Trespassers. No evidence remains of the level crossing and on the far side the track bed is lost to gardens, houses and a cutting.

Oxton to Lauder station 6½ miles

Head up Station Road, turn left on The Loan and continue south on the minor road towards Burnfoot and a small parking area from where the track bed is regained. A wooden signpost indicates the spot, Oxton ½ mile and Lauder 4½ miles. Despite numerous gates, what follows is a lovely peaceful stretch between fields, later with a burn on the right.

A modest embankment leads to the former level crossing at the road by Shielfield, where back the way is signed Oxton 1½ miles. The indicated way onwards to Lauder entails turning right on the road to Midburn and then a long, steep detour. (Although the track bed goes straight on, it appears that access agreement was not reached with the local farmer concerning the first ½ mile south of the Shielfield road.)

Turn left at Midburn for the climb to a telecommunications mast and an entrance to Blackchester farm. Turn left by a white locomotive sign, Lauder 3 miles. The steep lane descends to a cottage, above which is a tree-covered hill fort; an ancient settlement which also has a Roman provenance. Continue as signed on a grassy farm way to regain the track bed (map ref 513506) by another gate and one of several white posts denoting the HP Gas Pipeline. There is also a signpost, Lauder 4.2 km.....strange the change to metric.

The track bed and the A68, now seen and heard, gradually converge towards Lauder. Reach the Trabrown road where back-the-way is signposted, Old Railway Walk 700m and Oxton 3 miles. The sign pointing ahead, Lauder 1½ miles, shows double the real distance for the railway walker, because Lauder station lay at the northern end of the town. Traverse a modest cutting—it may be wet and in part churned up by horses but there is a good path on the right. A curving embankment over the Harry Burn then leads over a short rise to the site of the station.

A lane curves right, past the Scottish Fire and Rescue station, to a road signposted, Lauder and Railway walk Oxton 4 miles. Turn left into the industrial area that has subsumed all evidence of the station and buildings. What a sad end to the track bed walk from Fountainhall!

It is then a short stroll into the town, where the Southern Upland Way intersects the A68. Lauder developed at a guarded crossing of the Leader Water. The defences were enlarged in 1672 to become Thirlestane Castle, today a country house set in extensive parklands.

Galashiels - Selkirk

Following the 6-mile single-track branch may be a disappointment, in that most of the track bed has been lost to A7 realignment and Selkirk`s modern industrial estate. Nevertheless, it is possible to walk the distance, following the Gala Water, the Tweed and the Ettrick Water, albeit partly by the A7 roadside path, and the regular bus service gives a speedy return.

There were two intermediate stations, Abbotsford Ferry (Boleside) and Lindean: the ferry (a rowing boat guided by elevated wires) gave connection across the Tweed to Sir Walter Scott`s Abbotsford House.

Map 73 is required.

Starting from today`s Galashiels station on the restored Borders Railway, a path leads to the south bank of the Gala Water, which the branch crossed after diverging from the Waverley line. Of this bridge only an abutment remains. Curve southwards via Galafoot Road, or Winston Road, to a roundabout, pass under the high A6091 concrete bridge and continue on Boleside Road; a designated cycle route which briefly coincides with the Southern Upland Way. This, the prettiest part of the walk, goes through a lovely wooded area between the Tweed and a short stretch of identifiable track bed in a cutting.

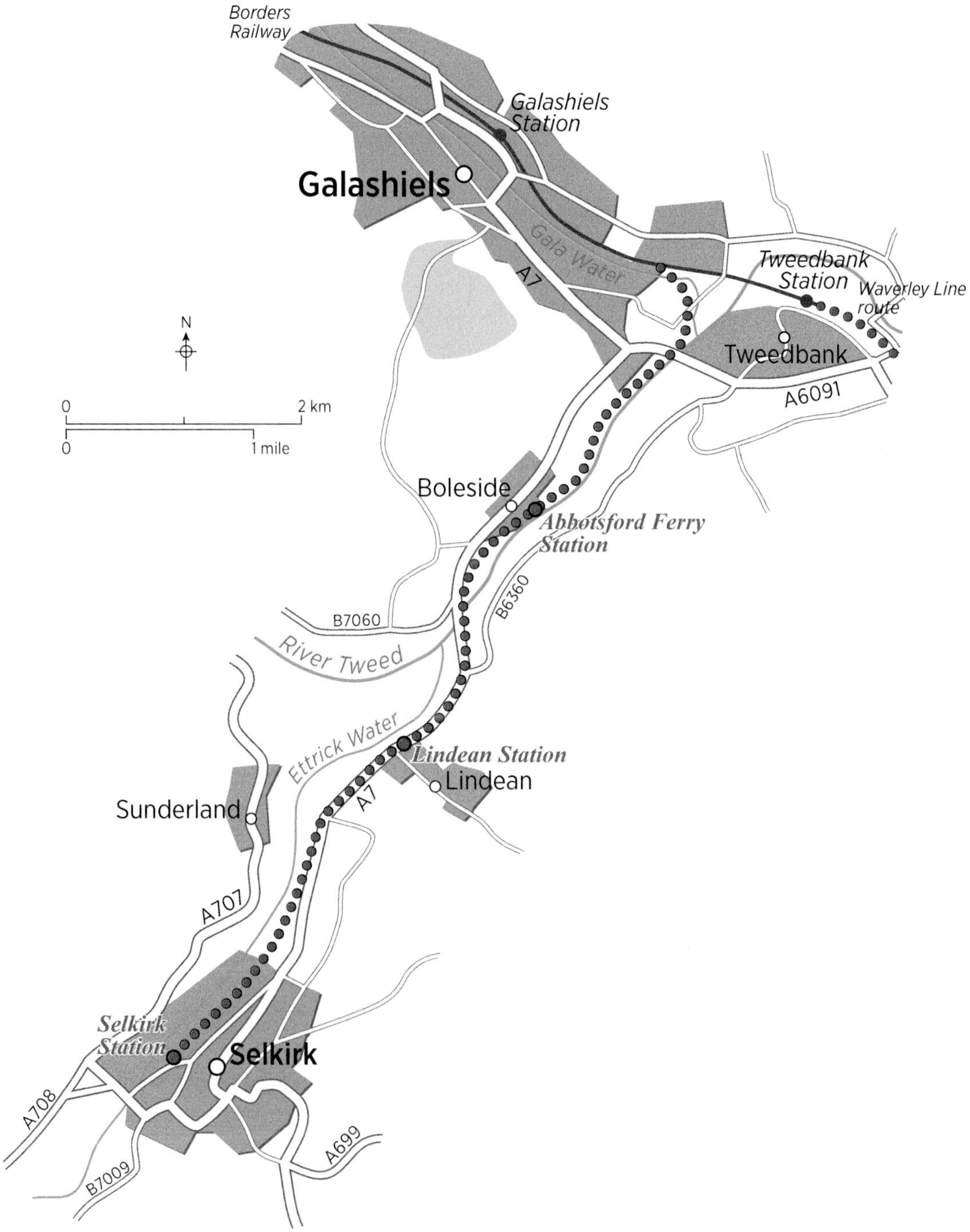

Borders Railway
Galashiels Station
Galashiels
Gala Water
A7
Tweedbank Station
Waverley Line route
Tweedbank
A6091
N
0
2 km
0
1 mile
Boleside
Abbotsford Ferry Station
B6360
B7060
River Tweed
Ettrick Water
Lindean Station
Lindean
Sunderland
A7
A707
Selkirk Station
Selkirk
A708
B7009
A699

Selkirk, in late LNER/early BR condition *Alsop*

Just downstream of the Tweed/Ettrick confluence, Boleside is reputedly one of the best Tweed fishing beats. No evidence remains of the station, though a plaque denotes the ferry landing.

A tarmac way leads to the A7 path. Cross with care to the junction with the B7060. From the mid-point of the new concrete bridge spanning the Tweed, there can be seen, just upstream, remnants of the masonry piers that carried the railway bridge. Beyond is the stone-arched bridge of the old A7. Use this to cross the Tweed then keep to the old road, crossing the Ettrick to reach Lindean. On the left is Station Cottage, the only evidence of the vanished station.

Follow today`s A7 roadside pavement/cycle way to regain part of the old A7 by Bridgeheugh Cottage, whence the riverside path (at one point recognisable as track bed) ends at Selkirk`s post-railway industrial units which occupy the low ground on the east bank of the Ettrick (note the new flood prevention works). A short straight stretch of track bed/tarmac path leads to the southern end of the industrial area, where lades brought water-power to the town`s numerous woollen mills. Beyond the still extant mill buildings is the site of Selkirk station, completely demolished in 1971. Only Station Road, the Station Hotel and the now derelict Station Road Post Office indicate the site....a sad ending to an otherwise enjoyable walk.

Tweedbank-Riccarton Junction-Kershopefoot

The long-awaited 30-mile Borders Railway from Edinburgh to Tweedbank, the longest domestic railway to be built in Britain for more than a century. Passengers have also been attracted by steam-hauled excursions featuring the now-retired Class A4 Union of South Africa and other preserved locomotives. Working hard on the climb to Falahill and

Melrose: NBR directors' special;
Class M 4-4-0 218 *Lynn*

then coasting downhill by the Gala Water, they recall the railway scene of 60 years ago; nostalgic reminders of sooty travel days. Though not fully appreciated from the train, two historic listed bridges are crossed; the single-arch Glenesk Viaduct over the North Esk, and the 23-arch Newbattle Viaduct, one of Scotland`s oldest railway bridges, over the South Esk.

Cost was trimmed by adopting single-track, with 9½ miles of "dynamic loops", more flexible in operation than traditional passing-places. In practice, margins have proved tight, causing delays; and special trains, save on Sundays, demand timetable alterations. To keep the service competitive, a reliable, less-than-60-minutes journey is essential. Nevertheless, early passenger numbers (commuters and shoppers heading north, tourists going south) have exceeded expectations and towns served by the line have seen significant economic benefits.

Railway enthusiasts cling to the hope that one day the line to Hawick, and perhaps on to Carlisle, will be re-instated. Indeed, at the time of writing the Scottish Government has instigated a study of road, rail and public transport in the Borders region. Under that review the Borders Railway could be extended. There would be major construction problems, for example at Melrose, St Boswells and Hawick, and the case for extending the line south from Hawick gets weaker by the mile. But one never knows, so, in the meantime, get walking!

The 46 miles from Tweedbank to the Border at Kershopefoot offer a fascinating and varied traverse that could be tackled over a long weekend.

A number of maps are required; 73, 79, 80, 85 and 86.

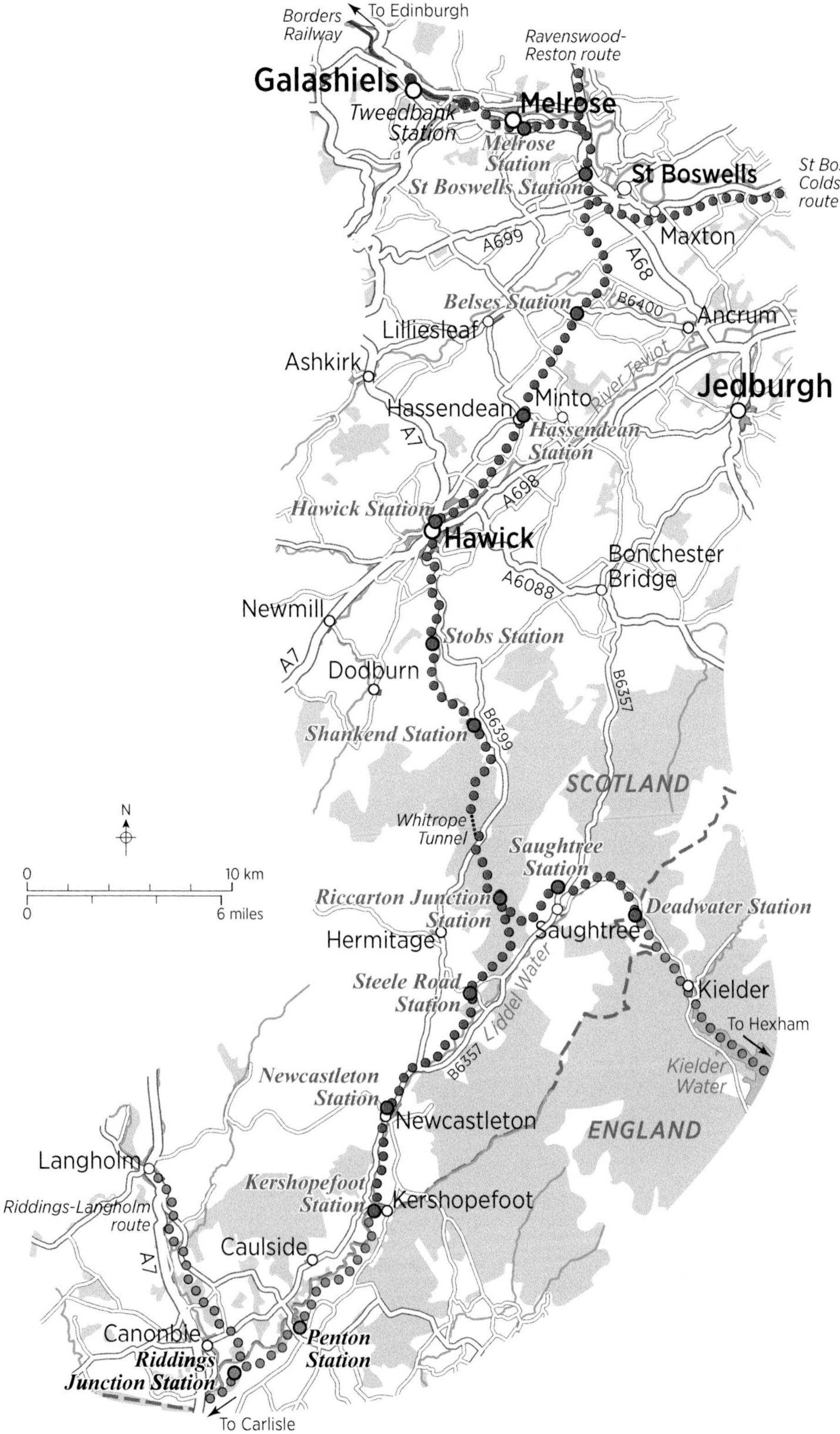

To Edinburgh
Borders Railway
Ravenswood-Reston route
Galashiels
Melrose
Tweedbank Station
Melrose Station
St Boswells Station
St Boswells
St Boswells-Coldstream route
Maxton
A699
A68
B6400
Belses Station
Ancrum
Lilliesleaf
Ashkirk
River Teviot
Jedburgh
Minto
Hassendean
Hassendean Station
A7
A698
Hawick Station
Hawick
Bonchester Bridge
A6088
Newmill
Stobs Station
Dodburn
B6357
B6399
Shankend Station
SCOTLAND
Whitrope Tunnel
N
0
10 km
0
6 miles
Saughtree Station
Riccarton Junction Station
Deadwater Station
Hermitage
Saughtree
Steele Road Station
Liddel Water
Kielder
To Hexham
Kielder Water
Newcastleton Station
Newcastleton
ENGLAND
Langholm
Riddings-Langholm route
Kershopefoot Station
Kershopefoot
Caulside
Canonbie
Penton Station
Riddings Junction Station
To Carlisle

Tweedbank to Melrose station 1¾ miles

Lacking station buildings and toilets, Tweedbank, mid-way between Galashiels and Melrose, is little more than a two-platform terminus.

From the station, follow the track bed tarmac path for ½ mile before a small industrial site blocks the way. Turn left to the road to Gattonside, but do not cross the Tweed. Curve right by the road-side pavement, cross with care the old Melrose road and follow the winding grassy paths through Darnick Community Woodland. Near the south end of the wood, seek out the site of a rail crash in 1921. A buffer from the debris, found during tree planting in 1998, is now displayed here.

From the south end of the wood, cross to the minor Ley Road which follows the old railway parallel to the A6091, the Melrose bypass. When the track bed is lost to houses, continue to the end of Ley Road. Turn right to the former railway bridge from where a tarmac path goes left, paralleling and then joining the track bed. Cross a road, after which a straight stretch leads to Melrose station, still signed but curiously bisected lengthwise, so that only the wooden frontage and part of the platform edge of the southbound line remain; a grassy strip separates them from the noisy A6091. All trace of the northbound line has gone. Descend by the central steps to admire the imposing station building. This listed structure, overlooking the town, contains a nursery, restaurant and offices.

Melrose to St Boswells station 3½ miles

From Melrose, the railway took a lovely curving route round the Eildon Hills, the most distinctive landmark in the Borders. Although much of the track bed has been lost (A6091 realignments; A68 Newtown St Boswells bypass; and in places ploughed up) it is possible with a few detours to have a most agreeable stroll.

Descend by Palma Place and turn right (eastwards) via Scotts Place to High Road, noting the high walls to the right behind which the track bed is partly subsumed by the A6091. Later, on the right, note Tait`s Well, once a public water trough.

When High Road curves right to the A6091, continue on Dean Road (the road to Newstead) for some 150 yards and then cross to the well-defined track bed path. Cross a bridge and slant right then left. On the first slant is an old railway sign for Newstead, a short-lived station closed in 1852. Later, Doctor`s Walk, a grassy path used by horse riders, adjoins the track bed.

Where the path divides, continue straight on then curve left (the track bed ahead is overgrown) and so by gravel path to the east end of Newstead, believed to be the oldest inhabited village in Scotland. Nearby is an informative plaque about the Roman fort of Trimontium, the place of the three (Eildon) Hills. (The Trimontium Walk from Melrose to Leaderfoot and back, weekly in summer, is recommended.)

Continue east on the Newstead road, then turn left just beyond the speed de-restriction sign. A gate leads to a mapped path; this hedge-lined, gently rising, grassy way is the old Roman road, mapped further south as Dere Street. Close your eyes and listen to the sound of tramping Roman legionaries. To the south the track bed has been ploughed up. From

a Trimontium notice board can be seen the impressive Leaderfoot Viaduct (refer to St Boswells—Reston Walk).

Continue to the stone parapets of the high bridge over the Reston line. Turn right and descend by a track (noting a path on the right which leads back to the Reston line) and so to the site of Ravenswood Junction. Continue past Broomhill farm and cross the A6091 to the far-side embankment from where a path leads to a gate by the A68. (At this point is an old wooden sign, back the way, Newstead and Trimontium.)

On the far side of the A68 is the driveway to Old Melrose House. Immediately on the right, the track bed is regained via the River Tweed path, and so to a cutting, at the far end of which is a bridge, from where steps on the right lead to a grassy lane. (Beyond the bridge, the track bed, a tarmac lane, cuts through a private golf course and is then lost to landscaping.) Turn right on the grassy lane, by the edge of the golf course, to reach the A68.

Follow with care the roadside grassy verge which improves to a broader strip opposite the bypass road to Newtown St Boswalls. Continue by the A68 for 100 yards then cross with care to the west side. 50 yards beyond a stepped brick path to houses is a secluded gate, from where a grassy path, following a line of telegraph poles, rejoins the track bed. This leads by industrial premises to a large car park and the site of St Boswells station, in its day an important exchange point where two lesser lines diverged. First simply Newtown Junction, it was briefly Newton St Boswells. (The village of St Boswells is two miles distant.)

The principal station building stood on the B6398 underbridge. Only the far side abutment remains; pink painted and festooned with hanging flower baskets in summer. Uphill from the abutment, steps lead to Bowden Road (behind are the Borders Regional Council offices). Immediately turn left to an industrial estate and left again to discover the decaying small engine shed which once served the lines to Reston and Kelso. The water tank, on its stone tower, survives too. (Hawick was the principal intermediate locomotive depot on the Waverley Route.)

St Boswells to Belses station 5 miles

From the engine shed, a winding path on the embankment leads to a bridge over a minor road, beyond which the track bed is lost to fields. Turn right on a narrow path between field and road, then left on a broad lane to regain the track bed; follow it to the site of Kelso Junction of which no obvious trace remains.

The track bed is then lost but the A699 can be reached by a pleasing diversion to the west. Retrace steps to the broad lane and turn left on the minor road; follow it to a woodland strip, map ref 573310. The mapped path, signed, Selkirk road (the A699), winds through the woodland, descends to a wooden bridge and then undulates on a grassy way by the edge of the wood to reach the A699.

St Boswells: departing Kelso local, headed by ex-NBR Class D30 4-4-0 'Wandering Willie' carrying BR number 62440; main line train at south platform; unidentified ex-LNER J39 0-6-0 on shed *Lynn*

Head east for some 100 yards to map ref 579305. To continue south to Thornielaw, descend by the Waverley line sign to a small cutting. The first section of track bed, strewn with cones, gives a wonderful stroll through a coniferous plantation. Beyond a high bridge it becomes, though used by tractors, a lovely tree-lined way in open country. A modest embankment leads to the Thornielaw underbridge beyond which there are good views back to the Eildons.

Thornielee, map ref 577293, was the location of Charlesfield Halt, opened during World War Two (1942) for workers at the nearby munitions factory, now the Charlesfield industrial estate. At peak production the factory employed 1300 people and made one million bombs per month. After the war the site became a Royal Navy Armaments Depot. The last recorded use of the Halt was in 1959.

Disappointment lies ahead. When farm and track bed part company, the latter becomes first wet and then impassable. There is no easy detour on foot, and it is advisable to arrange transport from Thornielaw to the minor road at Greenend.

In-filling and a wet, overgrown cutting are deterrents south of Greenend. This time, however, there is an easy minor road detour by Longnewton to the road bridge, map ref 588271. At first glance the cutting may seem too overgrown, but do persevere on the short prickly descent (harder in summer) to regain the track bed. The reward is a 1½-mile walk to Belses station.

The going quickly improves by a deep and broad grassy cutting and then through open country. Reach the high 4-arched viaduct over the Ale Water; difficult to photograph from below. At the next bridge, over the road to Belses Mill farm, only the embankments on either side remain. A grassy path to the left leads down to the road and on the far side, signed Thistle Doo, a gravel lane regains the track bed.

Beyond a grassy cutting, wet in places, and once clear of some trees, there are panoramic views to the right. Pass under an accommodation bridge from where the wettish traverse is best done at times of minimum growth. The going then improves. Just beyond a spoil heap, with New Belses farm to the right, reach a lane. Turn left to the B6400, thus avoiding the short embankment which ends abruptly at the vanished underbridge.

On the far side is Belses station, built primarily to service the 4-mile distant village of Ancrum. Although the track bed between the platforms has been filled in, all the station buildings have been conserved within a beautiful landscaped private garden. I was privileged to be shown around by the owner. The meaning of "Belses" is mysterious, though a farmer`s wife liked the suggestion "fields of gold".

Belses to Hassendean station 3½ miles

Assuredly in the top ten must-do station-to-station walks, this section is unique in that both stations have been preserved, with the manicured Belses contrasting with the back-to-what-it-was Hassendean.

Belses *Lynn*

A very short detour is required to reach the minor road south of Belses station. Uphill beyond the former railway bridge, map ref 573249, a track on the right gives access to the lovely track bed, used by tractors and horses, and from a modest embankment there are superb views back to the Eildons. Beyond an overgrown dump (likely to be removed I was told), Standhill farm is soon reached. On my visit a massive glasshouse, the largest of its kind ever constructed in the Borders, was being erected to produce organic tomatoes.

Beyond the bridge carrying the road from Greenhouse, the track bed is lost, but an enjoyable ¾ mile detour fills the gap. Continue by a minor road to the entrance to Minto Kames farm. The tarmac farm road passes by a lovely pond with mandatory heron and then crosses the line of the track bed, on the far side of which is pasture and more ponds. Stay with this road to where the tarmac ends and turn left to two gates, one blue one silver. Go through the latter and another silver gate to regain the track bed.

It is 1¼ miles to Hassendean. The straight track bed, used by tractors, goes under the Clockerhall bridge, passes by the Minto Hills (like miniature Eildons) and a large pond on the right, and then curves left through a wide expanse between trees. Farm tractors use this length and when their road curves left, continue on the grassy track bed to a gate and a Hawick paths sign. Beyond can be seen a white footbridge over the railway line.... Hassendean station is close to hand.

A gate leads to the cultivated strip of track bed which straddles the minor road.

Acquired from Buccleuch Estates, the overgrown site and derelict buildings have been saved and restored by architect Tom Pyemont who has made it his life's work. Decorated

with railway artefacts, the main station building has been filled with furniture of the 1930s. What a superb job Tom has done, and he is pleased to show visitors around what is his private property.

The original footbridge has been restored and it is a pleasure to climb from one side to the other. The restored station is like a step back in time; even the clock is stopped at 6.35pm. Beyond the footbridge, the track bed continues to what was once the loading bank and the signalman`s cottage (now a self catering facility).

Hassendean to Hawick station 5 miles

Descend from the station by the metal steps, turn right on a track which merges with the track bed and pass by a farm steading. A slightly overgrown section leads under the road bridge, but thereafter the broad embankment, with fields on either side, gives a delightful stroll through open country. Reach a cutting, discouragingly overgrown but not impassable, some ½ mile before Hassendeanburn farm. Then a fence bars the way. Cross with care (it had an electric top wire on my day!) and easy walking, on a grassy embankment, resumes.

The track bed, used by farm vehicles, leads to a cutting, again deep, overgrown and very wet. However, there is an easy bypass. Follow the tractor way which slants up on the right, parallel to the cutting, and so to Hassendeanburn farm steading. Turn left, then right, to regain the track bed. (In the Hassendean direction, it can be followed for a short distance.)

Pass under a metal latticed footbridge and then a stone-arched bridge, at first partly hidden, carrying the minor road from Clarilaw. Ahead is one of the longest straight stretches of the Waverley Line; 1½ miles, much in a shallow but dry cutting, used by farm vehicles; and this eventually slants to a minor road. Thereafter comes a mixture of road, meandering path through a wooded strip, and track bed in a minor cutting. Note the Waverley Way signpost.

Leave the overgrown track bed by the 30mph speed restriction sign and so into Hawick where almost all evidence, of what was an extensive layout, has been lost. Continue by road, pass a rugby pitch and then slant left, at which point it may be best to ask for directions to Hamilton Road. From there the line of the track bed is regained.

Hawick: two ex-NBR Class J36 0-6-0s and BR Standard 2 2-6-0 on shed (locomotive depot occupies original station site); main line south curves through later station to Teviot viaduct; 0-6-0s employed as banking engines for climb to Whitrope. *Lynn*

Pass by the remains of a short bridge, and then a long straight gentle descent, following the line of the track bed, leads through a pleasant park and by a cemetery to Teviotdale Leisure Centre and car park....the site of Hawick`s second station. (When the Border Union Railway was completed the original site became Hawick`s goods yard and engine depot.)

The station was completely demolished by 1974 and by 1982 the site had been levelled for the Leisure Centre. The area was also heavily landscaped in 2000 to accommodate the nearby Waverley Road Bridge. The only railway remnant is the edge of a loading bay.

Hawick to Stobs station 6 miles

From the Leisure Centre walk past the roundabout to reach the Teviot, perhaps first going upstream to the North Bridge, built in 1832 but pedestrianised when the Waverley Bridge was opened to carry the A7. Downstream are the foundations of the 42ft-high six-arch Teviot Viaduct which carried the railway. It was demolished in 1975.

Cross the Teviot and go up Mart Street, once the line of a high embankment. Pass the bus station on the left. At the top end, beyond a roundabout, is a very obvious grassy knoll with a bench, reached by steps to the left. Here the track bed, signposted Waverley Walk, resumes.

A lovely tree-lined path leads under Melgund Place Bridge, recently renovated.

The bridge over Loch Park Road is no more, but simply follow the detour path on the left. Ahead, the track bed has been lost amidst a small industrial estate, but a path on the

Stobs: troop train; NBR rigid wheel-base stock predominates *Lynn*

right is well signposted. Turn left to regain the gently rising grassy track bed which gives superb views of the town. Pass on the left a cemetery. The embankment, now a lovely tree-lined secret way, passes under a bridge to enter an attractive, deep, rocky (but dry) cutting.

Then the cutting comes to an abrupt end.....the majestic six-arch viaduct that spanned the B6399 and the Slitrig Water was demolished in 1982, allegedly after children were frequently caught throwing ballast onto the road below. Wooden steps, installed in 2009, ease access to the river where there is little evidence of this fine bridge.

Although the Slitrig Water could be splashed over during a dry summer spell, the more prudent way to reach the far bank is to walk south on the B6399 to just before the speed derestriction sign. Follow the track on the right to a Hawick Paths sign, Whitlaw Wood and Flex, from where a slanting path leads to the high embankment. (To view the river, descend steep but easy grassy slopes, interspersed with small trees.) Then follow a most pleasant dry cutting, curving southward; and so to the next grassy embankment, high above fields on the right and the B6399 below. Approaching the Flex road, a wet and partially blocked cutting is best detoured by following the east rim.

A farm track dips to cross the outflow from Williestruther Loch and then rises to re-join the true line of the old railway just before a stone-arched accommodation bridge. Ignore the farm track slanting to the right and stay with the now grassy but walkable track bed which gently rises on an embankment. Cross a roughish part above the outflow from Acreknowe Reservoir. The next section is well used as an exercise area for horses.

Once through a tree-lined grassy cutting, there are good views all round of the underbridge that spanned the Acreknowe road; only the side buttresses remain. Retreat a few yards, descend east and then turn right on the road to regain the track bed, now a broad grassy way which, to the left, overlooks the B6399. Above Woodfoot Pig Farm, the track bed curves towards the tall, 4-arched Barns Viaduct; an impressive listed structure but not readily photographed. (The south side is barricaded by brushwood, beyond which is the site of Stobs station.)

On the north side, paths lead steeply down to a minor road passing under the viaduct. (Downhill to the left is Stobs First World War Memorial, where the B6399 curves left to cross the Slitrig Water.) Head uphill, passing by two private lanes which lead to the Stobs stationmaster's house, now private. The garden, with more brushwood fencing on the south, embraces the old station platforms, linked by a distinctly unsafe footbridge, and features a brick-built ticket office. (Stobs Station was very briefly called Barns.)

Stobs to Shankend station 3 miles

Continue uphill on the road for some 700 yards to a cattle grid and a small parking area where, to the west, can be seen traces of Stobs Military Camp, which became known as Scotland`s Aldershot.

Because of the similarity of the terrain to the South African veldt where the Boer War had just been fought, in 1902 the War Office acquired the 10,000 acre Stobs Estate to be used as a

Shankend viaduct; EE Type 2 on BR officers' special *Lynn*

summer training ground for the British Army, both regular troops and 'territorials`. During 1914-18 as many as 15,000 men were stationed there. In addition, Stobs held 6,000 German POWs—a smaller number were detained during World War Two—housed in asbestos and corrugated iron huts. After 1945 the Camp saw use as a staging post in the repatriation of Polish troops. It finally closed in 1956, with the land sold in 1959. Photographic evidence survives of the large railway traffic generated over the years by Stobs Camp in its various guises.

From the parking area, follow a track as far as a metal gate, and then slant down southerly through a reedy field and cross a fence to meet a broad farm gate with a horse-rider friendly gate alongside. Beyond lies the track bed. (Heading north back to the station means traversing a deep, overgrown and well-nigh impassable cutting.)

Back to the track bed walk! Though slightly overgrown, the tree-fringed track bed/path offers a pleasantly secluded start. Later can be seen the rooftops of Stobs House, a castle-style mini-mansion. Nearby, hidden to view is the site of the remains of Stobs Castle.

Further south, the track bed becomes an easy broad grassy way, used by farm vehicles. Heed the warning sign at a small bridge high above a stream. Keep to the middle to avoid holes.

Reach Shankend Viaduct, a Grade 11 listed building, which spans the Langside Burn. 199 yards long, 60ft high and with 15 masonry arches, it is arguably the outstanding structure on the Waverley line and all the more impressive when seen from the B6399. Despite the sign, "Highways Act 1980. The British Railways Board hereby gives notice that this way is not dedicated to the public" it is possible to walk over the grassy way. At the far end is the private garden of Shankend Station Cottage.

Return to the north end of the viaduct. Follow the track on the right, turn left under the viaduct and regain the track bed by a slanting lane, once the approach to the station. A platform edge and the brick-built signal box have been painstakingly preserved.

Shankend to Whitrope Siding 4½ miles

A mile further south, a forestry track comes in from the right to subsume the track bed. When it diverges to the left, a wooden signpost (with a white locomotive symbol) reads… Hawick 9 miles, Kielder 20 miles. Keep to the track bed and continue SW through a large clearing and into the partly felled forest. A steady rise leads to a cutting and the northern portal of Whitrope Tunnel, at 1,208 yards the fourth longest in Scotland, beyond which is Whitrope summit (1,006 feet). The ruling gradient from Hawick through Stobs and Shankend is 1 in 75, easing to 1 in 96 as the summit is neared.

The tunnel is double-barricaded at both ends so does this mean the end of the walk? No. Return to the beginning of the cutting. Seek out on the east side, by a turning area, the start of an old pony track; at first reduced to a narrow path partly hidden by overgrown broom, but better defined as it zigzags to the high ground of Sandy Edge. The tunnel`s much deteriorated ventilation shafts should not be approached, so be very circumspect if straying from the pony track.

It is an easy pony track descent to Whitrope Siding, close to the highest point of the line ....and here are locomotives and carriages stranded by a red phone box! These belong to the Waverley Route Heritage Association (WRHA). Their storage yard contains old artefacts too, for example a dismantled lattice footbridge on its side. Walk back on the re-laid rails to have a closer examination of the southern tunnel portal and the immediate rock fall.

(By the WRHA car park, the pony track is signposted as the way to the summit and tunnel shafts. Given that the north end of the pony track is not immediately seen, a south/north traverse may be preferred.)

Whitrope to Riccarton Junction 2½ miles

Beyond the short Whitrope "restoration", the track bed curves left over the B6399 and into a deep, broad but dry cutting, at the far end of which is a padlocked metal barrier, easily circumvented. Now into open country, the track bed, a well-worn way used for forestry purposes, gives superb walking, curving through forest and cleared ground on the 1½ miles down to Riccarton Junction, the most isolated station on the Waverley line. (The ruling gradient, continuing down Liddesdale, is again 1 in 75.)

Final years; diesel-hauled train at bleak Whitrope Siding *Lynn*

This lonely railway centre of some 30 cottages and more than 100 residents entirely lacked road access until 1963 when a forestry track-cum-bridle road was constructed from the B6399, a mile north of Whitrope Siding.

Riccarton village had its own school, post office and grocery shop—the last a branch of Hawick Co-operative Society, situated in one of the station buildings. "Church trains" ran to Hawick and Newcastleton on alternate Sundays. From 1963, when the school closed, pupils travelled daily. Medical emergencies required a special train or light engine, as did occasional incidents involving the police. After 1956 the dead-end of the closed Border Counties Railway was used to store redundant locomotives.

Today Riccarton can be considered one of Scotland`s lost villages. Its sole resident occupies the former school building (which included the headmaster`s house), as a private dwelling on the hillside overlooking the station. The stationmaster`s house, grandest in the village as befitted his status, is a roofless shell. Few traces of the railway layout survive and the ash dumps, an inescapable adjunct of steam traction, have been excavated to make breeze blocks.

Having obtained a long lease from the Forestry Commission, Friends of Riccarton Junction, a voluntary organisation formed in 1997, aim to restore as much of the station as possible. An island platform has been partially reclaimed and a station sign erected, but little else appears to have been achieved and nature is slowly reclaiming the ground.

Riccarton was always a stark and utilitarian place amid bare bleak moorland slopes, little relieved by the uniform planting, post-1945, of Sitka spruce. Modern forestry practice, which avoids blanket coverage, has effected some improvement, but the station area remains desolate.

As with many a mining village in a scarred landscape, beautification seems neither appropriate nor practical. Nevertheless, Riccarton has become one of those must-visit

locations, not just for nostalgic railway enthusiasts but for walkers and a few cyclists. There is nowhere else quite like it.

Riccarton to Steele Road station 4 miles

The actual junction is marked by a wooden signpost, whence the line slants right for its sinuous ongoing descent into Liddesdale (while the Border Counties line heads SE, then eastwards, to Saughtree and Deadwater).

The heavily forested uplands are left behind; recent felling has transformed the general outlook and gives open views down to Riccarton farm and the B6357. Subsequent to the closure of the line, the Forestry Commission obtained additional access from the south along the track bed.

Where forest road and track bed part company (the former joins the B6357) follow the latter; it continues straight ahead as a lovely sheep grazed grassy way on the open southern slopes of Arnton Fell. The track bed, also used as a farm way, gradually curves left on a small embankment and so to a bridge over the minor road that links the B6399 and the B 6357.

Beyond a gate, a garden leads to the site of Steele Road station. However, respecting the curtilage, descend right before the gate, go under the bridge and then turn right to pass by the handful of named artisan cottages, including the blacksmith`s forge. Continue through the charming hamlet which grew up around the railway station. At the southern end, beyond the red phone box, the station entrance lane is signed Station House.

Snowplough and engine crews, January 1963; ex-LMS 0-6-0 carrying BR number 44081 heads a team of three locomotives *Lynn*

Steele Road (called Netherhope until 1900) was a modest intermediate station, serving a scattering of farms and houses. I was privileged to be shown round the stationmaster`s house, booking office and waiting room; all now converted into a charming private residence. "NBR 1891" is inscribed above the main door. (The North British Company, c.1890, came under increasing pressure from the Board of Trade to improve its earlier stations and railwaymen`s houses.)

The remains of a platform can still be identified. Beyond today`s garden is a wet and overgrown area, so retrace steps to the road.

Steele Road to Newcastleton station 4½ miles

Head south for a short distance and turn right on the forestry track to regain the track bed which is also used for vehicle access. Gradually curving right, it gives a charming walk with open views overlooking the Liddel Water...a grand and easy start on the continuing descent to Hermitage Viaduct.

After clearing the trees, the track bed traverses tussocky moorland and then an embankment. Beyond a new gate--note the footpath sign—it becomes a well-grazed sheep way. A very broad and grassy embankment leads to a shallow reedy cutting, best traversed to the left. Three muddy cattle-feeding areas lie ahead. Detour the first and second by a stile and marked path (on the left). The third, with no side path, may be challenging. Beyond the latter, a lovely grassy curving cutting (at the end of which is another path and stile) leads back to sheep country and a grassy way.

A high embankment above the Riever Sike leads to a small cutting occupied by covered sheep pens. (If unable to walk through, follow the rim on the left.) At the far end, by an overbridge, track bed and farm road coincide. When the latter slants down to the B6357, continue on the track bed, curving right to a plantation. On my day ballast was being excavated.

Beyond a fence, a weaving path on the gradually developing tree-studded embankment eases the way. It is possible to cut left through the trees and then, by a broader path, reach the B6357; however the intrepid walker will continue to the end of the embankment and descend steep grassy slopes to the B6399.

Once spanning the Hermitage Water and road, sadly Hermitage viaduct has been demolished.

Hermitage Castle, which lies four miles upstream, has a reputation, both from its history and its appearance, as one of the most sinister and atmospheric in Scotland. Only a few weeks after the birth of her son, James VI, Mary Queen of Scots made her notorious journey on horseback from Jedburgh to visit the wounded Bothwell there. They were to marry shortly after the murder of her second husband, Lord Darnley, although Bothwell was amongst the conspirators. How far the Queen was coerced remains debatable.

Walk to the B6357/B6399 junction and turn right across the twin-arched road bridge. A few yards on the right, two stone pillars mark the entrance drive to Redheugh. Go along the drive for some 100 yards to the end of the wall on the left. At this point a few stone blocks, remnants of the viaduct, can be seen by the river edge.

Turn left on a developing path and climb to the towering tree-studded Sandholm embankment which falls gradually to the south. The B6357 on the left comes into view and from there the track bed runs adjacent to the road all the way to Newcastleton. A lovely mossy way leads past a house, riding arena and paddock, and later crosses a gravel track, the access to Roan Farm.

Newcastleton: goods coasting downhill from Whitrope; ex-LNER Class K3 2-6-0, carrying BR number 61909 *Lynn*

On the almost level final stretch, the ill-drained track bed/grassy way may be wet in places. The going improves beyond a farm track until reaching a cutting and overbridge. (This is close by the 30mph restriction sign and a "Welcome to Newcastleton".) The cutting is prone to flooding. On my admittedly wet day the area under the bridge was more akin to a small pond, only crossed with the help of a plank of wood.

Very much a local walk, safer and easier than the road, it is then a charming stroll into the village; the only significant centre of population between Hawick and Carlisle. Passing by houses on the left, reach Lidalia Caravan Park (the address is Old Station Yard). Access is for residents only; however, a tree-lined way on the left leads to the road and North Square.

The lay out immediately calls out "planned village", and indeed it was—built on a gridiron pattern by the 3rd Duke of Buccleuch in 1793 as a handloom weaving centre on ground formerly known as Copshaw Holm. The previous settlement of Castleton, now mapped as Old Castleton, lies further up the Liddel valley. Earthworks mark the site of Liddel Castle, perched 100ft above a bend in the Liddel Water.

Turn right and continue to Douglas Square (but not as far as South Square), then right again on Langholm Street to what was, and is, the obvious level crossing point. The now private residence on the corner, Station House, bears a sign, crossing, no gates. A short stretch of track bed runs back to the caravan park fence. Beyond the station house, two large modern residences occupy the station area.

Heavy southbound trains were assisted from Hawick up to Whitrope and likewise, as required, north from Newcastleton on the 700ft climb to Whitrope. The last train before closure in 1969 was held up for more than an hour by a sit down protest on the line by the villagers.

Do spend a little while in Newcastleton, winner of Beautiful Scotland, best small village 2010 and 2014—famous too for its traditional music festival in July and for the Holm Show

in August. Well worth a visit, the Liddesdale Heritage Centre and Museum is in the former Congregational Church at the southern end of the village. The display includes memorabilia of the Waverley Route.

Newcastleton to Kershopefoot station 3 miles

And now for the final three miles of the Scottish section of the Waverley line to Kershopefoot, which, by just a few yards, is in Cumberland.

The grassy way/track bed, a charming local walk, leaves the site of Newcastleton station, passes by a plantation on the right and houses on the left, and slants towards the B6357. Road realignment has removed all trace of the overbridge and on the far side the line of the track bed has been lost. Continue south by road to a large hut on the right. On the opposite side of the road, a gate leads to a hidden, though mapped, narrow cutting; oft water-logged and more like a small canal in appearance.

Follow the right rim of the "canal." From the far end, the grassy track bed rises on an embankment to the Liddel Water....and traces of the stone piers of yet another demolished viaduct. The far side grassy embankment is well seen but the broad river enforces a detour.

Border signs *Lynn*

Return to the south end of the village and cross the road bridge over the river. Turn sharp right and then right again at the wooden post, Kershopefoot and Mangerton, to reach Riverview Holiday Park which has subsumed the track bed. From the far end of the park, follow the farm road that crosses a now filled in cutting. Turn left at Mangerton to regain the track bed, at this stage a broad gravel way used by the farm.

Pass by a cottage and a wooden signpost, indicating straight on to Kershopefoot, back the way to Newcastleton, and right to Mangerton Tower. Despite being just a few yards distant from the track bed, the remains of the 16th century tower house, now reduced to its lowest storey, are easy to miss. Seeming little more than a grassy bump at first, they are more impressive when viewed from the far side. Masonry repairs have been completed to the satisfaction of the Armstrong Clan Association, though visitors enter the site at their own risk.

With wooded slopes to the left and flat fields to the right, the tree-fringed track bed, wet and muddy on my day, follows the at times close-by meandering Liddel Water. Beyond a stone-arched overbridge, a modest embankment leads to the bridge over the Kershope Burn. For purists, the midpoint exactly marks the Anglo-Scottish border.

By the 16th century, the border was divided into six areas called Marches. At this point the Scottish Middle March meets the English West March. The Kershope Burn merges with

the Liddel Water to form the Scotland/England border as far as the confluence with the River Esk.

A few yards beyond the border is the road, on the left, to Kershopefoot village (the name Kershope comes from the Anglo-Saxon *coerse* meaning cress and *hop* secluded valley) with a Welcome to England road sign. To the right a triple-arched road bridge spans the Liddel Water.

Just over the road is Kershopefoot station of which little evidence remains. The stationmaster`s house stands in a hollow, its garden rising to the track bed where there are slight traces of a platform. Beside the house is a wooden hut, oddly bearing a sign, station house.

Though all the Border crossings were once distinctively marked, only two examples can be seen today—on the busy main lines north from Berwick and Carlisle.

Riccarton Junction-Deadwater (see Map on p.57)

Saughtree, the name means willow tree, was the only Scottish station on the Border Counties line, though the aptly named Deadwater station is/was only just in England. In 1908 in The Romance of Northumberland, A.G.Bradley wrote, "In a wild hollow.....the North Tyne springs from peat mosses, and on its way down lingers silently for a time in a rushy flat

Riccarton Junction: general view, looking to Whitrope; Border Counties line diverges right *Lynn*

known to the borderers as Deadwater, a name now embalmed in the timetables of the North British Railway". The outlook he found "wild, solitary and beautiful, incidentally disclosing a couple of farmhouses and perhaps twice as many cottages". Little has changed.

Having crossed the Tyne-Liddel watershed, the Border Counties line climbed through Saughtree to Riccarton Junction, whence the Waverley Route, heading south, descended into lower Liddesdale, a notorious haunt of the Border Reivers. (Refer to Tweedbank-Kershopefoot Walk.) Riccarton Junction faced north and there was little exchange traffic in the Carlisle direction. In effect, the Border Counties gave the North British Company a roundabout route to Newcastle, via Hawick and Hexham.

The following walks are described as starting from Riccarton. However, if starting from Deadwater, and continuing past Riccarton, and then to Whitrope Siding, be aware of the ambiguous wooden signpost at Riccarton. Slant left (WNW) for the pleasing 2½ mile walk northwards on the Waverley line track bed. It would be all too easy to go past Riccarton`s old school house and follow the broad forestry road which runs three miles to the B6399, one mile north of Whitrope.

Maps 79 and 80 are required.

Riccarton to Saughtree station 2½ miles

Forestry roads have subsumed both the old Waverley Route, south towards Steele Road station, and the Border Counties track bed, at first SE towards Saughtree. These diverge at the ambiguous wooden signpost.

The latter, a broad way, slants left, passes by a gravel quarry (on the left) and into a wide cutting. (The line was built `single` but sufficient land was taken for double track.) Reach an embankment high above a burn. Beyond the turning area for forestry vehicles, a good path follows the now grassy track bed, amid forest areas cleared or newly planted, which offer open views. Then enter a long, possibly wettish, grassy cutting; no problem for the well shod. The track bed, used for farm access, next traverses exposed, bleak moorland slopes. From its summit (870ft), near Riccarton, the old line ran down-hill almost all the way to Hexham, on a ruling gradient of 1 in 100.

Ahead, by a cluster of trees, is what may appear to be, on a misty day, a small but extant station..... Saughtree. This could be a Brigadoon moment, a step back in time.

Film goers of a certain age will remember the musical, Brigadoon, with Hollywood's idea of quaint Scottish villagers. "On a Scottish hunting trip, two Americans, Gene Kelly and Van Johnson, get lost in the woods and come upon the magical village of Brigadoon. Rising out of the Scottish mist only once every 100 years, and only for one day, Brigadoon is an enchanted place where life is simpler and easier, untouched by the worries of the modern world. When Gene Kelly falls for beautiful villager Cyd Charisse, he must decide whether to stay or to return to his life...."

Patiently waiting at the single platform, beside a signal, is a small locomotive, Meg of Saughtree (albeit diesel and not steam), with a guard`s van and two wagons. Everything, including the stone-built station house, is in pristine condition.

I am indebted to Geoff and Meg Mann for their detailed story of the station (which dates from 1862), its resurrection and its future possibilities. There never was a signal at Saughtree and the engine shed for Meg is a converted railwayman`s cottage, reached by a siding on the right. The former ticket office has been converted to a B&B (saughtreestation.weebly.com); a godsend for the railway enthusiast. The station house, extended and now a private dwelling, retains its charm.

The station was sold in 1987 (over 200 enquiries, many from railway enthusiasts). Though the project, as it presently stands, was completed in 2008, nevertheless work continues and any suggestions for "railway improvements" are always welcome. The owner`s son-in-law kindly let me explore the site. He told me that, had the owner been there at the time, he would have insisted in taking me for a ride on Meg of Saughtree---a third of a mile out and back trip. Bad timing on my part!

A mile south of the remote station, the hamlet of Saughtree lies by the B6357 at the Liddel Water/Dawston Burn confluence.

Though passenger trains were few (typically, three in each direction daily, and suspended for a time during World War Two), the Border Counties line generated a useful revenue (especially coal from the small, isolated Northumberland mines at Plashetts) and only became hopelessly uneconomic after 1945. Excursion traffic from Newcastle was another staple, almost to the end. The LNER Northern Belle (ancestor of all today`s "cruise trains") regularly ran south from Edinburgh via the Border Counties line—Abbotsford and Hexham were both on the day`s itinerary.

Saughtree to Deadwater station 3 miles

Leaving Saughtree, pass by a railway hut and on to the grassy embankment towards the B6357. In the 1960s, not a railway-friendly decade, Scottish Borders Council chose to demolish the 70ft high viaduct spanning the road and the Dawston Burn. Descend right to the station entrance road. Cross burn and road and climb to the far side embankment. The immediate cutting can be avoided on the right. From there the grazed grassy track bed, overlooking upper Liddesdale, leads through sheep country to a plantation, mapped but being clear-felled on my day.

A forestry track, coming in from the left, subsumes the track bed to give an easy stroll to Myredykes and the minor road to Kielder. Road realignment has erased all remains of the overbridge and the far side cutting is avoided by briefly following the track to Fairloans. Regain the track bed (signed Kielder), now a grassy way. To the right can be seen old lime

kilns. A short stretch through a plantation leads to open ground on the left and a wooden post that marks the Border.

Welcome to England! The former Border sign has been preserved in the National Railway Museum at York. From the post, a path on the left leads to the road where other signs confirm that Scotland has been left behind.

However, it ill behoves the walker to stop at the border. Deadwater station is only 500 yards distant on a good path. Reach a broader grassy way with remnants (on the right) of Thorlieshope siding which served a limestone quarry.

And so to Deadwater station. As with Saughtree, the isolated station had little traffic and full public facilities at this lonely spot only dated from 1880. The single platform survives and the station house, again stone-built, is now a private home.

In the 19th century the nearby Deadwater Well had a modest fame. In 1825 Eneas Mackenzie wrote that the well was much frequented by persons suffering from "cutaneous and scrophulous complaints and who receive great benefit by drinking the water and by using it as a warm bath". But by the 1860s, when the Border Counties line reached what was known as Deadwater Foot Crossing, patronage of the Bathing House was already in decline.

St Boswells-Reston

This 28-mile St Boswells (Ravenswood Junction)-Reston cross-country line, which had seven intermediate stations.....Earlston, Gordon, Greenlaw, Marchmont, Duns, Edrom and Chirnside, offers a most enjoyable and varied traverse and passes many a place of interest. In short, it is a must-do line.

A new two-hourly local rail service between Edinburgh and Berwick-upon-Tweed is projected for 2021, with the stations at East Linton and Reston restored or replaced. The walker who ventures from St Boswells, arriving at Reston before the station`s re-opening, will surely plan a second visit—this time by train.

Maps 74 and 67 are required.

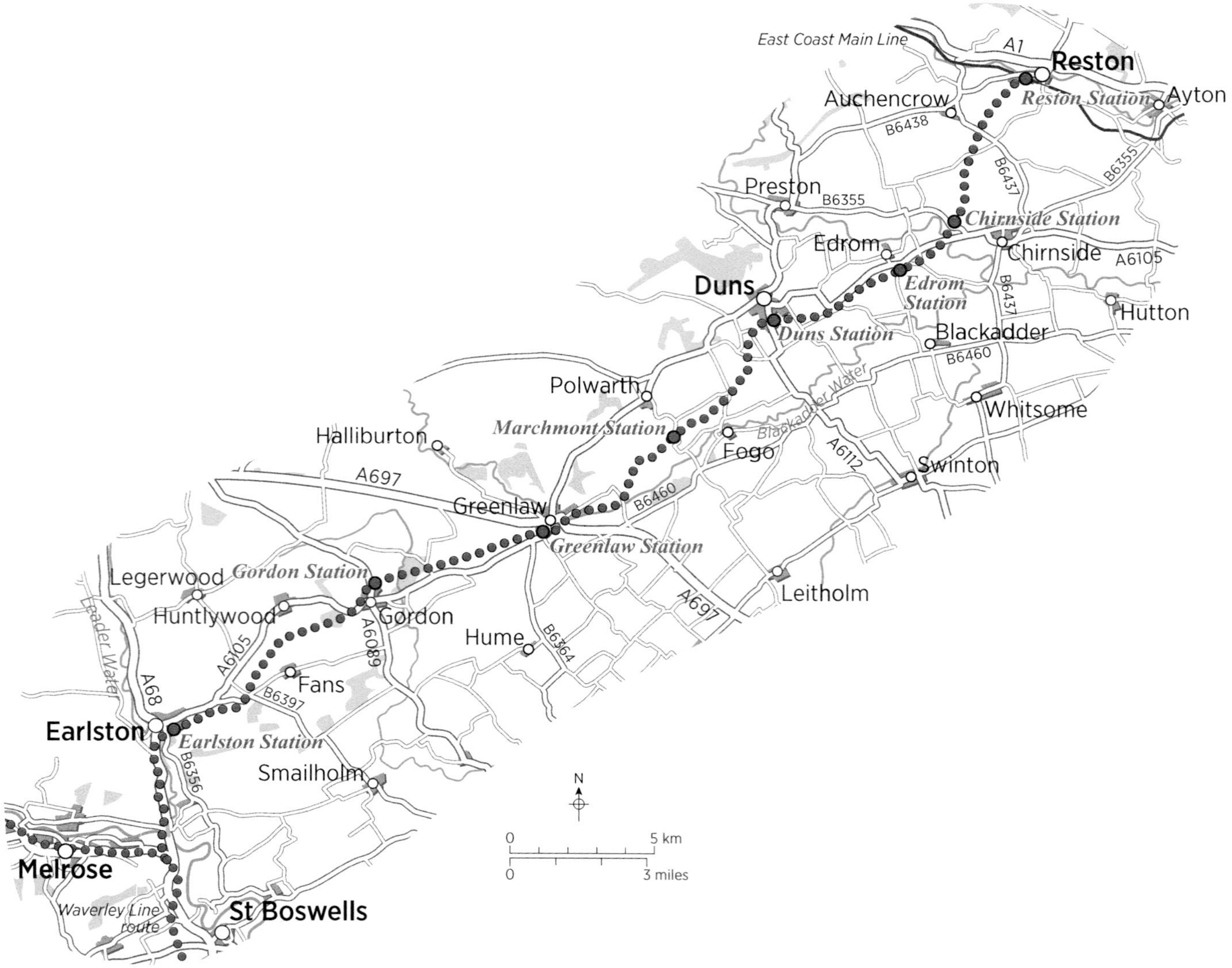
East Coast Main Line
A1
Reston
Reston Station
Ayton
Auchencrow
B6438
B6355
B6437
Preston
B6355
Chirnside Station
Chirnside
A6105
Edrom
Edrom Station
Duns
Duns Station
Hutton
B6437
Blackadder
B6460
Polwarth
Marchmont Station
Blackadder Water
Fogo
A6112
Whitsome
Swinton
Halliburton
A697
Greenlaw
B6460
Greenlaw Station
Leitholm
A697
Legerwood
Gordon Station
Huntlywood
Gordon
A6089
Hume
B6364
A6105
Leader Water
A68
Fans
B6397
Earlston
Earlston Station
B6356
Smailholm
N
0
5 km
0
3 miles
Melrose
Waverley Line route
St Boswells

Ravenswood Junction to Earlston 3 miles

The site of Ravenswood Junction (refer to Tweedbank—Kershopefoot Walk) lies just west of the A68/A6091 junction. Head north from Broomhill on a farm track, and where it slants uphill, follow the signed path on the left that leads into a cutting. The lovely track bed path passes under the triple-arched bridge (used on the walk from Newstead) and later into open country. On the left are a Trimontium notice board and a well-placed viewing platform.

Continue a few yards on the embankment to the Leaderfoot Viaduct, barred by a padlocked gate topped by barbed wire; insufficient to prevent the all too many suicides. (The drop to the water is 126ft.) However, this sombre aspect is offset by the sheer beauty of the much photographed graceful 19-arch viaduct that spans the Tweed. It survives as the line`s outstanding feature. Demolition was once threatened but the viaduct is now listed and in the keeping of Historic Scotland.

Return to the Trimontium notice board, cross the stile, descend by path and steps to an old road, and turn right. Walk under the viaduct to the three-arched Drygrange road bridge. Built in the late 18th century, the bridge was reconstructed in 1926. Further downstream is the ugly though utilitarian A68 bridge. The cluster of three bridges, railway viaduct included, has been dubbed "Tripontium".

Across the Drygrange bridge is the Gattonside road (B6360). A scramble to the northern end of the viaduct (also gated) is feasible, but not worth the effort. Much of the track bed onwards to Earlston has been subsumed by the A68 and the roadside grass verge is not recommended. Nevertheless, the delightful signposted Leader Water path, opened in 2010, comes to the rescue.

Turn right on the B6360 and pass under the A68 (Leader Water path, Earlston 3½ miles) and continue with care on the busy curving road. Turn left just before the bridge over the river (Earlston 3 miles). Well-signed throughout, the winding way gives a lovely secluded walk by the wooded west bank. Pass by the garden walls of Grange Hall nursing home and later cross the Water by a metal bridge. Follow the east bank "Speedys path" to reach the B6356 and enter Earlston at Station Brae. On the right is the entrance to Earlston industrial estate; the site of the station, but nothing has survived.

Earlston to Gordon station 5½ miles

Although the line of the dismantled railway is easily determined from the map, and there are physical remains (e.g. the abutments of the overbridge at Purvishaugh), the first two miles from Earlston High School towards Gordon offer little in the way of a walk.

The regular bus service is one option but that would mean missing the most hidden and exciting part of the old line....the traverse of Gordon Moss, the surviving remnant of a much larger moss. A birch woodland and a Site of Special Scientific Interest, it lies in a shallow depression, between the Fans road and the A6105, through which flows the Hareford Burn which later joins the Eden Water. Once the site of a loch, the Moss became a source of peat fuel and was divided into three areas to define local rights. Its eastern side is in the care of the Scottish Wildlife Trust.

Transport complications once resolved, the easiest approach to regain the track bed is from Fans, a hamlet south of the old railway. From there, an off-road walk, used by residents at Fans to get to Gordon, traverses the Moss. A broad farm lane descends to the track bed, the site of Fans Loanend siding.

Head left (west) on the tractor way/track bed to the end of a wooded strip, beyond which is pasture for cattle. Retrace steps and continue east. With another wooded strip and marshier ground by the Hareford Burn on the left, the grassy tractor way/track bed gives a dry traverse, well away from roads, and heads directly into the secret Moss. Take care where the tractor way, easily mistaken for the track bed, curves right. The true route is straight ahead!

Cross a broad wooden bridge on the left, then a railway culvert and so by a silver birch lined way towards more open terrain. Pass on the right a sign, Gordon Feuars Moss.

Gordon in NBR condition
Lens

All too soon, a gate leads to open pasture. To avoid any cattle, follow the Moss-side field edge to the A6105. On the far side are the ruined remains of Greenknowe Tower, a 16th century tower house occupied until the 1850s and now in the care of Historic Scotland. Follow the roadside verge, pass by the abutments of the A6105 underbridge (the embankments on either side offer little walking) and so to the west end of Gordon.

One approach to Gordon station is to head along Main Street (Cafe Obrigado has a picture of Gordon station) to the east end of the village and turn left onto Station Road, the A6089. However, a more pleasant approach is to turn left by the community hall and go along Manse Road. Turn left and right along Sunnybraes. At the end of the houses is a popular local walk, Edinburgh Road. Descend towards the overgrown track bed and so to a dry cutting; a secluded defile approaching the A6089 overbridge. Take the path on the right before the waterlogged area below the bridge.

On the far side, signed Old Station Yard Gordon, is a small industrial estate. In contrast to the obliteration at Earlston, Gordon station house (now extended) is a private residence and the surrounding garden has been landscaped. However, few other traces remain of the platforms and other buildings.

Gordon to Greenlaw station 4 miles

The grassy track bed beyond Gordon station is obvious, but there are many electric wires to be crossed and, on reaching Gordon Community Woodland, the bridge over the Eden Water is no more.

However, an attractive diversion is to hand. Head south on Station Road to Gordon Cottage and turn left on the mapped farm lane which passes by a large pond. Reach a wooden bridge over the Eden and immediately turn left. Follow the meandering path at the western end of the wood, adjacent to the burn, to regain the track bed which takes a lovely straight course through mixed woodland; an area extensively used by the local community. Rest a while at a charming pond. Curving left to avoid the boggy track bed, a woodland track reaches a car park by the East Gordon road junction.

The close-by overbridge has been filled in. On the south side is a metal gate (Beware of Bull) from where a track leads to a forlorn sign, BEWARE OF THE TRAINS. The track bed, at first a smooth tractor way, becomes grassier amid sheep and cattle pasture and heads directly for Greenlaw.

Greenlaw station was sited at the southern side of the village, and west of the boarded up track bed under the curving A6105. Apart from the station house, now a private dwelling, little evidence remains. The area has become a general storage yard for trailers and caravans.

Whilst in the area pay a visit to the Bedshiel Kaims; a sinuous 2½ mile ridge, 15m high in places, consisting of layers of sand and gravel deposited by glacial streams. To reach the esker, head north on the A6112, partly follow the Greenlawdean road but then continue NNW by a wooded strip and into moorland. Rather like a man-made flood protection bank, the ridge is a commanding viewpoint over the bleak surroundings and its pebbly soil, topped with sparse vegetation, is a delight to walk on.

Greenlaw to Marchmont station 4 miles

Leave Greenlaw station by the drive on the left and regain the track bed on the far side of the road by a sign...Football ground ½ mile. (If walking with a dog, make a circuit through the village.) Later slant right, following the yellow arrow signs. (If there is a bull nearby, retrace steps.) A modest grassy embankment leads to the A697 and another sign, back the way, walkers welcome but no dogs for your safety. On the left is the W.S. Happer memorial park, home to Greenlaw FC, and community woodland.

On the east side of the road is Crossings Cottage from where the track bed leads to the Blackadder Water....and another missing railway bridge. Retrace steps to the road bridge and turn right on Marchmont Crescent to a faded red gate on the right. Cross the grassy area to regain the track bed, and then pass through a sequence of gates and by the sewage treatment works. (Alternatively, continue by road to the speed de-restriction sign and turn right on a tarmac lane.)

What follows is a lovely track bed stroll. Passing by the prominent Blackadder Water footbridge, a local walk from Castle Mill, the track bed becomes a tractor way. On approaching the impressive road bridge over the Water, tractor way and track bed part company. The latter goes via an overgrown cutting to a filled in overbridge. Follow the tractor way to the road and head north to a gate. Descend to regain the track bed and pass by the remnants of a long platform, Lintmill siding.

During World War Two the siding served RAF Charterhall, used by NO 54 Operational Training Unit. Nowadays a private airstrip, the airfield lies 1½ miles to the south-east. At a triangle of roads opposite Charterhall Wood is a memorial to Flight Lieutenant Richard Hillary and Sergeant Kenneth Fison and the many others who lost their lives between May 1942 and May 1945.

Continue north over the grassy area, pass under an accommodation bridge, and cross a burn three times. What follows is a secret, seldom-used tractor way through slightly overgrown cuttings, and which eventually rises to a grassy embankment and more open country; all the while a beautiful sinuous route round the southern side of the hidden 390-acre gardens of Marchmont Estate.

A gentle descent leads to a minor road and Marchmont station—a public facility, though its main purpose was to serve Marchmont House. The station house is now a private residence and its landscaped garden occupies the track bed by the platform.

Marchmont House, at one time used by the Sue Ryder Foundation as a care home, has now reverted to private ownership. Stretching north-east from the house, the puzzling mapped feature is the Marchmont Great Avenue; at 1.3 miles, thought to be the longest in Scotland. Planted in 1726 with 10,000 Dutch elms, the avenue is still extant, albeit with other trees, and is best appreciated from the minor road that cuts through the avenue at right angles.

Marchmont to Duns station 3½ miles

Regain the track bed by the access lane of another new house. What follows is a lovely stroll by the Howe Burn to the minor road junction south of Ladyflat (map ref 773505). The deep and overgrown cutting leading to the overbridge is avoided by taking a tractor way curving to the left.

Overgrown and lost to fields and urbanisation, from this point the track bed approach to Duns comes to an end. However, from the junction head NE by road to Gavinton. From the south-east end of the village, take the minor road to where it crosses the Langton Burn by a cobbled ford. Too much water? Take off boots and paddle over! Go north past Langton Mill, and then turn right past Grueldykes to a mapped lane (map ref 782530). This grassy hedge-lined way leads to a relatively new tarmac path.

Turn right at the sign, Duns Scotus Way, Duns Park ¼ mile. Duns is the probable birthplace (1266) of John Duns Scotus, one of Europe`s great philosophers of the late middle ages who explored the conflict between faith and reason. A curving duckboard way leads on to a tarmac path, and so into the beautiful park to meet a fine statue of the man himself.

Another name associated with Duns is that of Jim Clark, the former world champion racing driver, born in the Fife village of Kilmany. The Jim Clark Memorial Room in Duns is open March to September.

This old market town, twinned with Zagan, Poland, has developed an unattractive sprawl, but its appeal will grow. Exit Duns Park and turn right (south) on what is Station Road. Just before the industrial estate, turn left on the old station entrance. The station house, now extended to a private residence, is attached to the former waiting room and ticket office. No artefacts remain, save a post-Grouping notice, once at the station entrance and now attached to the modernised station house. "The London & North Eastern Railway Company hereby gives warning to all persons not to trespass upon any of the railways, stations, works, land or property belonging to or worked by the Company. Trespassers are liable to a fine or imprisonment for every offence".

Duns to Edrom station 3 miles

Continue south and then turn left into the extensive Duns industrial estate. Where that road curves right, a gate on the left gives access to an excellent path (Crumstane 1 mile) —in fact the former railway. This quickly leads to a road where the missing railway bridge presents no problem. The track bed continues in grassier form on a modest embankment to a wooden gate and an old overbridge (on a minor road). A second gate leads to a small plantation beyond which the track bed, parallel to the A6105, has been ploughed up.

A short walk by the busy road to Brieryhill is now required, Walk past the (closed) south gates of Manderston House, "the supreme country house of Edwardian Scotland, the swan song of its era". It has, reputedly, the world`s only silver staircase.

At Brieryhill, a minor road spans a lengthy and overgrown cutting, which appears to bar progress; but a local walk, direct to Edrom station and rediscovering the track bed en route, comes to the rescue. Go down the right hand side of the cutting to a wooden gate and traverse the grassy strip by the field edge. Continue some distance to the end of the cutting, and at a farm track crossing point, regain the track bed.

A grassy way between fields leads to a lovely, tree-lined, long but modest embankment. Open parkland follows until, beyond a gate and a road overbridge, the station suddenly appears. The house by the platform was the ticket office and the small mounted clock is stuck at 9.40. There is also a sign on the wall, "Warning is hereby given to persons not to trespass upon the railway. Penalty not exceeding 40 shillings". The house at the top of the entrance drive was the stationmaster's, now a private dwelling.

Edrom to Chirnside station 1½ miles

Although the station-to-station walk is only 1½ miles, the short distance provides a fascinating route, enhanced by minor detours.

The track bed, at first a grassy tractor way, fringed by rosebay willowherb, draws the walker on for some 700 yards to an overgrown cutting. Slant left, follow the field edge and head through a plantation to reach the road bridge over the cutting. The track bed beyond is lost but another readily identifiable railway overbridge at Craigswalls points the way. This is reached by the triangle of minor roads to the east.

Cross the Craigswalls bridge and turn right, passing the farm storage area. Though there is no mapped track bed onwards to Chirnside, a path, again very much a local walk, covers virtually all the distance. Cross a small dump to gain a broad grassy strip by the field edge, leading to the southern end of the mapped wooded strip which hides a railway cutting....deep and wet in part, and best avoided. Continue by path on the east side of the wooded strip to reach a cluster of old and new houses. Avoiding them, the path curves right. Thereafter, turn left to the small metal gate that leads to a grassy area on the high bank of the Whiteadder Water, and so to the A6105.

A short distance to the left, the realigned road has subsumed the railway cutting, and on the far-side the track bed is now an entrance to the Chirnside Bridge Paper Mill. Originally constructed in 1842, the extensive plant, which occupies the low wedge of land by the Whiteadder Water between the A6105 and the B6355, is nowadays the Ahlstrom Chirnside Mill; a specialist paper plant which produces over 10% of the world's tea bags. There is no public access, and what remains of the five-arched Whiteadder railway viaduct cannot be discovered.

Cross the river at the utilitarian concrete David Hume Bridge, built in 1968 to replace the graceful old bridge just upstream. Scotland's David Hume, the outstanding philosopher of the European Enlightenment, lived nearby at Ninewells.

100 yards further on, to the left, is a marker post, signed David Hume Walk, and another, signed BBW (Borders Brains Walks). The other Berwickshire geniuses are James Hutton, Duns Scotus, James Small and Alexander Dow. There is also a green disc with a yellow arrow; the first of many to be encountered on the way to Reston.

The path, slanting left through a wooded area, follows the east bank of the Whiteadder to reach the B6355. A short detour west on the B6355 leads to the high padlocked gates at the northern end of the mill premises, where the tall brick chimney stack is obvious. There is no sign of the railway viaduct but on the right (north), a short stretch of track bed is recognisable.

Return by road to just west of a red telephone kiosk, from where a lane descends north to the stricken site of Chirnside station, nowadays a large farm storage area, mainly for potatoes. A short length of platform survives, together with the station house and an attached green wooden building. All have seen better days.

The hillside village of Chirnside lies one mile to the east.

Chirnside to Reston station 4 miles

Before the station house, turn left as signed, then right (but keeping left of a wooden fence). Beyond further traces of a platform, a farm gate leads to the field edge from where the track bed is regained. Despite a few detours, the onward walk to Reston is most satisfying. Green disc/yellow arrow signs are plentiful and successive finger posts show mileage to Auchencrow and Reston. The many gates are horse-rider and walker-friendly.

The track bed, parallel to the Billie Burn, reaches the road to Billiemains. The underbridge is no more, but steps lead down from the embankment. Across the road, beyond a gate, is the site of Billiemains Siding and then, just before the B6437 is reached, is the site of Auchencrow Siding. Both were on the western side of the line. No tangible evidence remains though the site of the latter is more obvious.

Continue to where the track bed is bridged by the minor road from Newmains. A path on the left leads to the bridge and a finger post, Reston 1.5km. Given the overgrown cutting, it may be tempting to follow the road into Reston. However, this means missing the completion of the railway walk at the merge point with the East Coast main line. Best done at times of low growth, persist through the cutting or descend steeply from the far side of the road.

The track bed, a lovely grassy way, then curves right to the remains of a platform. Keep to the right and turn left on a path through the plantation strip where the lines once converged. Then follow the field edge to reach the B6438. Head east through the village to a lane on the right, which leads to the site of Reston station. The station house is now a private residence.

On my travel through Duns I met a lady who told me that her mother, as a young girl, had worked at the station until being made redundant when the station closed (1964). Despite the passing years, the mother was minded to re-apply for her old job....or, at least, hope to be invited to the grand re-opening.

St Boswells-Roxburgh Junction-Carham

Diverging from the Waverley Route (refer to Tweedbank—Kershopefoot Walk) to follow the south bank of the Tweed, the line had seven intermediate stations in Scotland....Maxton, Rutherford, Roxburgh, Wallace Nick, Kelso, Sprouston and Carham, and five in England.... Sunilaws, Cornhill-on-Tweed, Twizel, Norham and Velvet Hall. In British Railways days, trains ran between Berwick-upon-Tweed and St Boswells, reversing at Tweedmouth. In 1948, when the Edinburgh-Berwick trunk line suffered exceptional wash-outs along the Eye valley, prestigious East Coast expresses were diverted via Kelso.

St Boswells-to-Carham, despite some diversions, offers a pleasing 15-mile route that could conceivably be done in one day.

Map 74 is required.

St Boswells (Kelso Junction) to Maxton station 2½ miles

Kelso Junction can no longer be identified with certainty (refer to Tweedbank—Kershopefoot Walk) and further south the track bed is lost. On the south-east side of the A699 overbridge (map ref 581306, and evident by the "bulge" in the road), a pond has formed in a lengthy cutting. Further on, approaching the road to the Charlesfield industrial estate, a short stretch of overgrown track bed is not worth the effort; and, to the east, a triangular wooded area necessitates a detour by road to the site of the railway bridge over the A68, identifiable only by its embankments.

At long last a walkable track bed lies ahead. On the far side of the road is a wooden marker post, signed left St Boswells Green and right, path link to Laretburn. Follow the latter to gain the far side embankment which gradually curves between the A68 and the A699. A lovely path on the embankment winds through a wooded strip. Then, beyond a fence, the track bed becomes a broad grassy way between fields. Easy walking continues, under an unmapped bridge and into a cutting. Reach the minor road SW of Maxton. Beneath the overbridge there may be a deep pond. If less than suitably shod, follow the right-hand field edge to regain the track bed for an easy stroll towards a white-painted house, the site of Maxton station.

(Alternatively, head north by the minor road, briefly follow the St Cuthbert`s Way and pass through the quiet village. St Cuthbert`s of Mackiston is first recorded in the reign of William the Lion, 1165-1214, though there may have been a church here before that date. Turn right at the NE end of the village to reach Maxton station.)

A broad well-kept lawn beside modern bungalows has obliterated both track bed and station. Of the level crossing, only one metal post and part of a deteriorating gate remain.

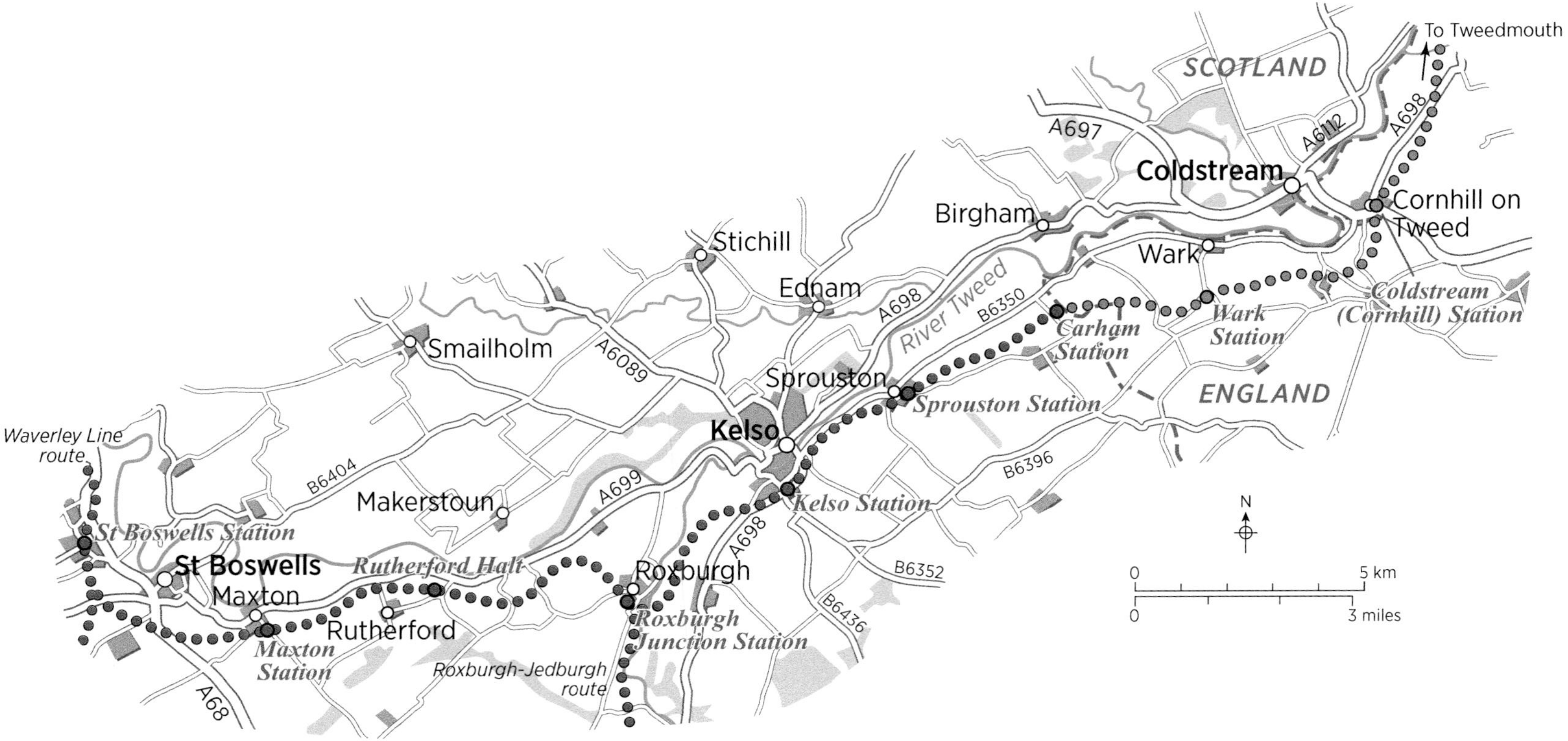

To Tweedmouth
SCOTLAND
ENGLAND
A697
A6112
A698
Coldstream
Cornhill on Tweed
Coldstream (Cornhill) Station
Birgham
Wark
Wark Station
Carham Station
Stichill
Ednam
A698
River Tweed
B6350
Smailholm
A6089
Sprouston
Sprouston Station
Kelso
Kelso Station
B6396
Waverley Line route
B6404
Makerstoun
A699
A698
St Boswells Station
St Boswells
Rutherford Halt
Roxburgh
B6352
B6436
Maxton
Rutherford
Maxton Station
Roxburgh Junction Station
Roxburgh-Jedburgh route
A68
N
0
5 km
0
3 miles

Steps lead down to the white-painted station house, now a private residence; a short section of attached wall is the only remnant of the waiting room and ticket office. The lady of the house, having very kindly showed me around the site, suggested I should help myself from the apple tree.

Maxton to Rutherford station 2½ miles

Maxton`s wide grassy lawn may give the impression that it will be an easy walk to Rutherford station. Alas, not so. Beyond the lawn, the immediate track bed is wet, overgrown and not recommended. Make the short detour via the A699 to Riddletonhill where the track bed to the west is used as a dump. However, to the east superb walking is on offer on the tree-lined embankment, also used by farm tractors.

Enter a small but mostly dry cutting; a lovely hidden way giving open views from time to time, especially back to the Eildons. A slightly rougher short stretch, best done at times of low growth, is made easier by a winding path. Passing over a minor road, the going then improves. The final approach via a cutting gives a secluded entrance to the site of Rutherford station, mapped as Rutherford Mains.

Beyond a lengthy platform edge on the right is an open area, now used as gardens and garages for a line of cottages. The most easterly cottage was the stationmaster`s house and the adjacent lower building has a sign, Rutherford Mains, Parish of Maxton. There is a red GR post box. On the left, at the end of the station site, is another short edge of platform propped up by decaying sleepers.

Rutherford to Roxburgh station 3 miles

A smooth track bed/tractor way goes under a bridge and then crosses a grassy way. Slightly further on, at map ref 670304, is an overbridge spanning a minor road, from where a gate leads to an embankment which gradually becomes more overgrown and ultimately well nigh impassable. Once over a burn, escape from the embankment by slanting down to the right. Follow the field edge tractor way until it turns left to intersect the old line of railway at a missing bridge and immediately regain the far side embankment. Fairly good walking then resumes to the road overbridge at map ref 677307.

(For an easier, albeit longer route from map ref 670304, slant north to the minor road, from where a detour by the A699 leads to the bridge at map ref 677307. Regain the track bed via a gate on the north side.)

It was by this bridge on my first visit that I met a number of men about to embark on a hare shooting outing. (The open season for brown hares is from 1st October to 31st January.) They advised that a series of water-logged and overgrown cuttings more often than not made the curving approach to Roxburgh impassable. Perhaps my visit was not opportune.

I returned another day to discover that the track bed, though wet and slightly overgrown in places, did not inhibit progress. Where also used by farm tractors, the track bed gives easy walking; a superb secret way, but still with no sight of Roxburgh. The track bed gradually curves round the slightly higher ground on the right. At a pile of ballast, before yet another cutting, make a detour to the right to the mapped lane, map ref 692312. Follow the lane over the deep overgrown cutting, whence Roxburgh at last appears. At the aptly named Over Roxburgh farm, join a minor road, which is followed south to a junction and then head west, signposted St Boswells, to the old railway bridge which spans the road.

Regain the track bed on the north side and head towards a broad grassy tree-studded open area—the site of Roxburgh station, with the stationmaster`s house, renovated as a private dwelling, almost hidden beyond the fence at the far end.

Retrace steps to the road junction and head south. On the right is the station entrance, identified as Roxburgh station house.

Continue south by road and turn right to Roxburgh Junction, where the Jedburgh branch diverged and both lines spanned the minor road. Both bridges have gone, leaving the high ivy-covered abutments as gigantic tombstones marking what once was. An impressive though sad sight!

On the north side of the road, two benches provide rest for nostalgic walkers. On the south side is today's West End car park, where two wooden marker posts identify the railway bifurcation—Jedburgh Line, caution cyclists, and Kelso 2½ miles.

In the 13th century Roxburgh was reputedly the fourth most populous and important town in Scotland. The remains of Roxburgh castle (Marchmount), which commanded the Tweed/Teviot confluence, are best visited from Kelso (below). In 1460 James II was killed at the siege of the castle, then in English hands, when a canon exploded near him. The castle was eventually demolished in 1551.

Roxburgh via Wallace Nick to Kelso station 2½ miles

A gate (no motor cycles) gives access to the continuing Kelso line. A broad path regains the track bed embankment which leads to the gently curving multi-arched 70ft high Teviot viaduct. Remedial works (British Railways Board and Railway Heritage Trust) have restored this attractive Grade B-listed structure, best appreciated and photographed from below. Cue for a worthwhile detour. On the south side, a slanting path on the left descends to the Teviot where the viaduct can be admired from both upstream and downstream.

After the problems on the approach to Roxburgh, what a joy it is to stroll on the track bed; a lovely public path, waymarked with yellow arrows within green discs. The very popular local walk gently eases away from the Teviot. Later pass by Maisondieu, once the site of an "asylum for pilgrims, the diseased and the indigent."

A missing railway bridge is detoured to the right, by which is a sign, back the way, Roxburgh 2 miles, and ahead Kelso, but with no mileage. After another longer but well-pathed detour on the right, regain the track bed which gently descends towards Wallacenick. No evidence remains of where the line halted in 1850 before being extended to Maxwellheugh, Kelso, a year later. The nearby farm and housing development is styled Wallaceneuk.

Eastward the A698 has subsumed the track bed, so follow the broad roadside pavement to a roundabout at the A698/ B6352 junction on the south bank of the Tweed, where Kelso station once stood. The extensive site has suffered the fate of many in being subsumed by an industrial estate, a supermarket and car park.

Over the years Kelso has expanded to the south bank and, with its bridge across the Tweed, remains an important transport hub—other lower river crossings, at St Boswells to the west and Coldstream to the east, are some distance away.

The "prettiest, if not the most romantic village in Scotland" was Sir Walter Scott's opinion at a time when Kelso lay predominately on the north bank of the Tweed. The town certainly has a most attractive square. Perhaps the largest of the Border Abbeys, Kelso's Benedictine foundation (1128) was severely damaged in 1545 during the Rough Wooing and little now remains. Floors Castle, the seat of the Dukes of Roxburgh, is said to be the largest inhabited house in Scotland. Nearby is Kelso Race Course.

Do remember to visit the traces—their scale is still impressive—of Roxburgh castle back along the Teviot. Follow the Borders Abbeys Way.

Kelso to Sprouston station 2½ miles

Continue along the A698. Where the road curves to cross the Tweed, look out for a signpost on the east side marked Sprouston, from where a path rises to regain the track bed. The next two miles, a lovely mixed route, offers a thoroughly enjoyable walk.

A bridge over a minor road is no more but the track bed is easily regained on the far side. A short distance further on is Mellendean, where there was the two-mile gap between the railway being built from the west and the one from the east. That gap was filled after a two-year standoff.

Beyond Mellendean, the track bed crosses another minor road close by the cluster of houses at Sprouston. Further east, the former station is easily identified by a ramp leading to a few yards of platform edge. At the B6350 there is a sign, Station House; externally the building, though modified, still shows its railway provenance. Beyond a tarmac lane new

bungalows and gardens occupy the station site. The nearest house has another short length of platform edge.

This was the only penetration of Scotland by an English company. By contrast, the North British Railway owned perhaps 100 route miles in England—if one adds up the Hexham, Morpeth, Rothbury, Silloth and Port Carlisle lines. "Like a Border reiver" says historian Hamilton-Ellis, the North British raided deep into Northumberland.

Sprouston to Carham station 2½ miles

Follow the track bed/lane by the new houses to reach the B6350 and turn left for a few yards where no evidence remains of the level crossing. Cross the road to a wooden signpost (Carham 2¼ miles) from where a winding path regains the track bed, over mapped as a lane, running straight ahead. This gives a superb walk.

Spanned by a minor road, a dampish cutting, used by horse riders, may be slightly churned up. However, after a farm track comes in from the right, the track bed, parallel to the Redden Burn, becomes a beautiful, broad and open way between fields.

Adjacent to the Redden Burn is a large pond to where mute swans migrate every year. My timing was impeccable. From May to the beginning of August, mute swans retreat to different sites in the Borders for their six-week annual moult which leaves them completely flightless. Indeed, owing to the rich feeding at the Tweed Estuary, Berwick-upon-Tweed is home to Britain's second largest such colony.

Pass by an old, 3-arched lime kiln built into the hillside, then through a new plantation to approach Carham station, which had a most unusual layout—its staggered platforms were intersected by the minor road from Nottylees. No station buildings have survived and there is no evidence of the level crossing, albeit the platforms, especially on the south side, are extant.

The nearby village of Carham, pronounced with a silent "h", lies by the south bank of the Tweed and just inside England.

There is still a short distance to walk. The national boundary, further east by a strip of trees, brings to an end what is a most enjoyable and interesting walk from St Boswells. No passport is required.

Between the Wars, the LNER installed Border signs and British Railways continued the practice. In this instance, east of Carham, the sign, now gone, was a post-Nationalisation addition. Did its earlier omission echo old inter-company frictions prior to the Grouping?

Roxburgh Junction-Jedburgh

Following the banks of the Teviot and the Jed Water, the Jedburgh branch is easily tackled as one continuous 7-mile stroll, mainly on track bed. It had three intermediate stations, Kirkbank, Nisbet and Jedfoot. The walk coincides in part with the 65-mile Borders Abbeys Way (BAW) developed around the four ruined abbeys at Melrose, Jedburgh, Kelso and Dryburgh. Though BAW signs are numerous, be aware that Way and track bed part company beyond Jedfoot.

There is a regular bus service between Kelso and Jedburgh, but this follows the A698, thus bypassing Roxburgh where the walk begins.

Map 74 is required.

Roxburgh to Kirkbank station 1½ miles

From the West End car park (refer to St Boswells-Carham Walk), a wooden signpost (Jedburgh Line, caution cyclists) identifies the path to an embankment. Thereafter the track bed becomes a slightly hidden way. During a wet spell, the initial part may be muddy, churned by many horse hooves. After one mile reach a lesser embankment and a gate, where steps to the left indicate where the BAW from Kelso joins the track bed. Thereafter the route to Jedfoot station is signposted with yellow arrows.

Some 800 yards later, descend by steps on the left; a detour necessitated by a missing overbridge--only the abutments remain. On the minor road from Ormiston Mains, turn left, then right. A black metal finger-sign, Kirkbank Stat, indicates the site of Kirkbank station. A short stretch of platform edge is still extant. Elsewhere the area has been levelled to incorporate the garden of the much modified station house, now a private dwelling.

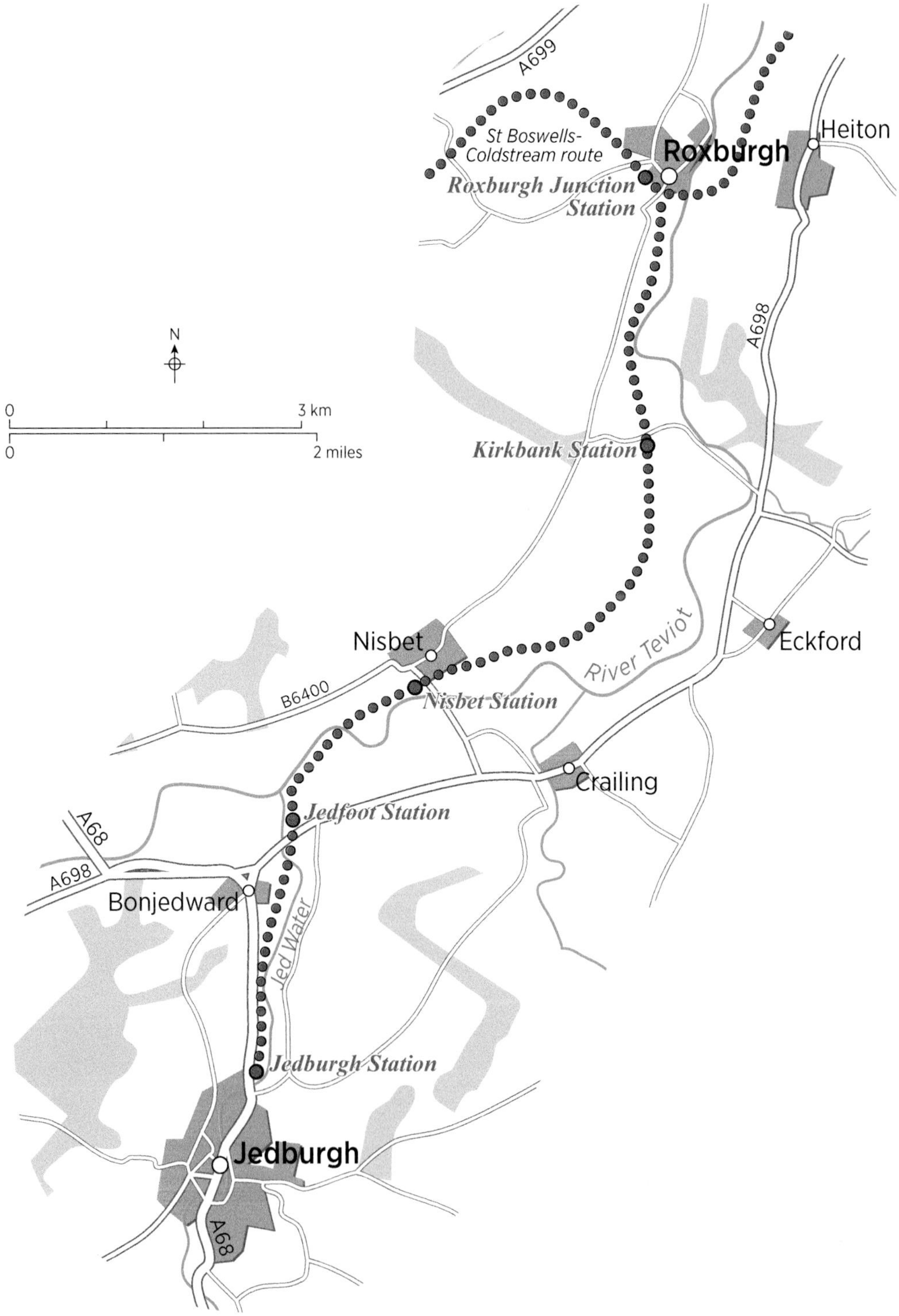

A699
St Boswells-
Coldstream route
Roxburgh
Heiton
Roxburgh Junction
Station
A698
N
0
3 km
0
2 miles
Kirkbank Station
Nisbet
Eckford
River Teviot
B6400
Nisbet Station
Crailing
Jedfoot Station
A68
A698
Bonjedward
Jed Water
Jedburgh Station
Jedburgh
A68

Kirkbank to Nisbet station 2¾ miles

This length may seem relatively nondescript; but curving SW, away from the meandering Teviot, there is fast and easy walking on a broad and firm surface—a reminder that the line was originally built double-track.

At a small cottage in ruins, the Kalemouth-Eckford path (signposted) goes left, passing Ormiston House. One mile further on, the path to Nisbet village (also signposted) goes right. Nisbet is a designated conservation area, almost unaltered since it was developed in the 19th century to house farm workers. Overlooking the village is Peniel Heugh, crowned by the 150ft tower built to commemorate the Battle of Waterloo. From afar, the monument looks like a lighthouse. Though there is no access to the top of the tower, its base is a commanding viewpoint and well worth a visit.

Continue to the B6400 and on the far side is the site of Nisbet station. There is no evidence of the level crossing gates. The station house (and post office) is now a private dwelling, Riverside Cottage. Attached to a fence is a street sign, EASTER ROAD—bizarre until I learned that the owner of the cottage is a fellow Hibernian aficionado, who, following the glorious 2016 Scottish Cup triumph, bought the old road sign for the bargain price of £35. Perhaps we shall meet up again at the real Easter Road.....

Nisbet`s single platform has been incorporated into the gardens of the handful of houses that occupy the former goods yard. The remains of a railway coach can just be detected in one garden a little further along the track bed.

Nisbet to Jedfoot station 1 mile

Resist the temptation to continue by track bed or farm lane to where the branch crossed the Teviot. Though abutments remain, the bridge is no more and fording the water to the south

Pier of Teviot viaduct (west of Nesbit) *Furnevel*

bank is not recommended. Cross the Teviot by the B6400 iron box-girder bridge and follow the path on the south bank flood embankment, as does the BAW—but make a slight detour to examine the remains of the rail bridge. Later regain the track bed and so to Jedfoot station, on the north side of the A698 and close to the confluence of the Jed Water and the Teviot.

Were it not for the road identifying the location, the station site would be easily missed. Only a short stretch of the stone platform edge is apparent on closer examination. Trees and bushes have proliferated and the station building has been demolished. Close to the eroded banks of the Jed Water is a line of decayed vertically-set sleepers.

Jedfoot station was never a grand affair but these remnants make wistful viewing.

Jedfoot to Jedburgh station 1¾ miles

Cross the A698 and say goodbye to the BAW which takes a higher route by the slopes of Mount Ulston to reach Jedburgh. It is also a quick hello and goodbye to the St Cuthbert's Way (intersected on the St Boswells-Carham Walk at Maxton) and the Roman Road, Dere Street (encountered east of Melrose on the Waverley line), both of which cross the Jed Water. Slant right to the short stretch of embankment that leads to the Jed railway bridge, of which only a circular pier and the south side parapet survive.

Retrace steps and cross the Jedfoot bridge, erected in 1970 to replace the old stone-arched structure. Carefully walk by the grass verge to enter Bonjedward. From the A698/A68 junction, use the A68 roadside pavement and after 500 yards turn left on the tarmac road to Bonjedward Mill Farm through which the railway ran. Turn right and cross a burn to regain the track bed for the final mile. A lovely secluded stretch becomes a pleasant broad way between the A68 and Riverside Park, home to Jed-Forest rugby club.

And so to Jedburgh station—or rather to the industrial estate which has entirely subsumed the station site. It may be a slightly sad end to the 7-mile walk but do continue to the attractive town centre square. (Given the ¾ mile uphill, a connecting bus service was provided for railway passengers.)

Although ravaged by the English in 1545, Jedburgh abbey in ruins is an impressive sight. The ancient ball game, Jethart Hand-ba, supposedly derives from the men of Jedburgh playing with the heads of English soldiers. The frequent incursions of the late middle ages were repelled with the Jeddart staff, a long pole tipped by a metal hook.

Fife, Clackmannan, Kinross

1 Cameron Bridge-Leven
2 Leven-St Andrews
3 St Andrews-Wormit
4 Glenburnie Junction-St Fort Junction
5 Kinross-Milnathort-Mawcarse Juntion-Glenfarg
6 Mawcarse Junction-Ladybank
7 Kinross-Alloa
8 Dunfermline-Alloa
9 Cambus-Alva
10 Markinch-Leslie

Fife, Clackmannan, Kinross

The amalgamations of the mid-1860s gave the North British Railway an almost total control of Fife, Clackmannanshire and Kinross-shire. Two decades later the Caledonian Company would reach across the Forth from Larbert to Alloa (1885). Though not without allies (who looked to counter the North British monopoly), they failed to expand eastwards towards the Fife coalfield and the holiday-cum-fishing towns of the Fife Coast.

In the earliest years of railway development, a direct line north from Edinburgh (i.e. from the Fife side of the Queensferry narrows) found some support – one ambitious scheme aimed at the Highlands via Glen Devon and Crieff. However the superiority of the longer but unbroken route via Stirling to Perth was soon acknowledged. The Edinburgh, Perth &

Burntisland: Edinburgh, Perth & Dundee Railway terminus (Granton Ferry interchange), in NBR condition *Lynn*

Dundee Company (formed in 1848 when the Edinburgh & Northern and the Edinburgh, Leith & Granton amalgamated) made the best of the (not inconsiderable) regional business available. They were resigned, however, to a poor share of long-distance traffic. Burntisland-Tayport constituted the Edinburgh, Perth & Dundee's main line; their Perth connection ran from Ladybank via Bridge of Earn. The original route to Dundee began with the short rail journey from Edinburgh's Canal Street Station to Granton (the Canal Street-Scotland Street incline was cable-operated) and ended with a still shorter stint from Broughty Ferry to Dundee East, with reversal at Broughty Junction on the Dundee & Arbroath Railway. The water-breaks of Forth and Tay (Granton-Burntisland and Tayport-Broughty Ferry) were enterprisingly overcome. For goods and mineral traffic the company introduced pioneering train ferries, designed to overcome tidal variation, while passengers crossed by conventional, well-appointed paddle-steamers. Nevertheless, it was altogether a weary experience, some three hours end-to-end at best.

The Edinburgh & Glasgow Company came to control and eventually absorb the Stirling & Dunfermline Railway, completed in 1852, and so penetrated into Fife. (From Stirling the Forth & Clyde Junction Railway provided a tenuous connection westward by Balloch to the Edinburgh & Glasgow's Clydeside lines.) North British dominance between Forth and Tay was not established until both the Edinburgh Perth & Dundee (1862) and the Edinburgh & Glasgow (1865) had been captured. (The Forth & Clyde Junction, nominally independent until the Grouping, was leased by the enlarged North British.) The Stirling & Dunfermline added an early branch to Tillicoultry; their Alva branch (from Cambus) opened in 1863, under Edinburgh & Glasgow ownership. Connection was made with the Thornton-Cowdenbeath-Dunfermline branch, built by the Edinburgh & Northern (1849). This created a Thornton-Dunfermline-Alloa-Stirling through route, on which coal traffic predominated; but Dunfermline's access to Burntisland and the Granton ferry remained circuitous.

At the very end of their independent existence, the Edinburgh & Glasgow Company obtained powers for a Ratho-Queensferry branch and thereby a prospective share in any future scheme to bridge the Queensferry Narrows. This threatened the North British, who had just united with Edinburgh, Perth & Dundee; but the eventual result was to hasten North British-Edinburgh & Glasgow amalgamation, after which the branch was built. A corresponding rail connection on the north shore, from Dunfermline to North Queensferry, waited till 1877. Meanwhile the south-of-Forth connection to Granton had been modified to run from what had become Edinburgh's General Station (Waverley in the making) via Abbeyhill.

The Leuchars-St. Andrews branch opened in 1852. The East of Fife Railway (soon reconstituted as the Leven & East of Fife) made stop-go progress from Thornton Junction to Leven (1854), Kilconquhar (1857) and Anstruther, reached in 1863. Formal North British take-over of both these lines waited until 1877. The Anstruther & St. Andrews Railway

St Monans: flower bed, local women and bare-foot boy; train formed of NBR rigid wheel-base stock save for one bogie vehicle *Lynn*

remained nominally independent until 1897: running via Boarhills (1883), it had its own station at St. Andrews (1887) until the North British added a short link through the town, thus completing a round-the-coast loop from Thornton to Leuchars.

Among other subsidiaries of the Edinburgh, Perth & Dundee Company, the Fife & Kinross Railway (from Ladybank) opened in 1858. The Kinross-shire Railway, from Lumphinnans on the Dunfermline-Thornton line to Kinross was ready two years later, and a joint station at Kinross was established in 1861. The Devon Valley Railway which fulfilled the original aim of the Fife & Kinross promoters – a Ladybank-Tillicoultry route and thereby another link with Stirling – was finished by the North British in 1871. During the short life of the first Tay Bridge (1878-9), a through Glasgow-Dundee service ran via Stirling, Alloa and Kinross (with reversal at Ladybank), by exercise of North British running powers over the Caledonian Railway between Greenhill and Stirling. In conjunction with the Bridge came new construction from Leuchars to Wormit, while the Newport Railway (1879), inevitably a North British client though with its own identity until 1900, joined Wormit with Tayport. This, together with the northernmost section of the original Burntisland-Tayport main line, created a secondary loop for local traffic between Dundee and Leuchars, worked from the new Dundee Tay Bridge station.

Events north-of-Tay lie outwith this Chapter but merit a summary. The rival Scottish Central route through Stirling and Perth, in the Caledonian Company's hands from 1865, terminated at Dundee West. When in 1866 the Caledonian Company absorbed the Scottish North Eastern Railway and won their way to Aberdeen, a North British interest in through traffic was preserved by placing the Dundee & Arbroath Railway under joint Caledonian-North British ownership. New construction added a purely North British link from Arbroath through Montrose to the Caledonian (Scottish North Eastern) main line at Kinnaber. From Tay Bridge Station the Dock Street tunnel connected with the Dundee & Arbroath. Through running between Perth and Arbroath was achieved by a Caledonian connection with the northern approach line to the Bridge. For the North British, all these developments were blighted by the collapse of their showpiece "longest viaduct in the world", whereupon the old service via Tayport and its ferry was abruptly restored.

With the new Tay Bridge in place (1887) and the Forth Bridge completed (1890) the North British became a serious competitor for all the traffic of the North, through Perth and through Dundee. The company's English partners benefitted too – a reward for their large contribution to the Forth crossing. But refashioning what were now unbroken trunk routes was no small task, and their piecemeal nature remained obvious. New construction to complement the Forth Bridge included the Lothians approaches from Saughton and Winchburgh; the Fife approach from Inverkeithing; an Inverkeithing-Burntisland connection; and a cut-off over Glen Farg from Mawcarse on the Fife & Kinross line to Bridge of Earn. In addition the existing North Queensferry (Inverkeithing)-Dunfermline and Dunfermline-Kinross lines were upgraded and made double-track throughout; this entailed a short cut-off at Cowdenbeath. By virtue of the old Midland Company's input, the Forth Bridge Railway survived the Grouping, in the shared ownership of LNER and LMS (cf. the Midland Anglo-Scottish route via Carlisle).

Fife's lesser lines included the Dunfermline & Charlestown Railway, which had begun as an 18th century wagon way to the Forth; in the longer term it served to round off a secondary route between Alloa and Dunfermline – the Kincardine Railway of 1893. This was conceived as something of a block line, in case the Caledonian resumed their attacks east from Alloa; completion to Elbowend Junction south of Dunfermline came only in 1906, after powers were obtained under the Light Railways Act. The Newburgh & North Fife Railway, by which the North British Company finally occupied that district, was another latecomer (1909); it ran from Newburgh on the old Ladybank-Perth route to a triangular junction at St Fort on the Leuchars-Tay Bridge line of 1878, creating a scarcely-needed additional route between Perth and Dundee, in competition with the Caledonian line on the Tay's north bank. (The Leuchars fork at St Fort, which permitted through working between Perth and St Andrews, lasted only a few years.). Here again was North British *over*-insurance – it had become increasingly unlikely that the Caledonian might yet probe into Fife. The North Fife line would remain independent on paper until absorbed into the LNER in 1923.

Coal and freight apart (and discounting many mineral lines – not to mention the private Wemyss Railway), the relatively dense rail network of Fife, Clackmannan and Kinross together with the great bridges made possible passenger workings now little remembered. Some endured into the 1960s – for example, the scheduled trains and summer excursions from Edinburgh and Glasgow to all the coast towns, or the year-round secondary service between Glasgow and Perth, via Larbert, Alloa and the Devon Valley to Kinross and thence

Alloa: in BR condition; ex-CR 0-4-4T *Lynn*

over Glen Farg. Unusual provision over the years included winter specials to Kinross, for bonspiels on Loch Leven.

It follows that the litany of closures is long and bleak, accelerating in the 1960s when completion of the Forth and Tay road bridges reinforced the gloomy verdict of the Beeching Report. Passenger services on the North Fife line ended in 1951, with complete closure by 1964. Leuchars-Tayport closed in 1956. The Wormit-Tayport branch, cut back to Newport East in 1966, was closed entirely from 1969. The Fife Coast line succumbed in 1965-6, though Thornton-Leven and Leuchars-St Andrews lasted three more years. In 1950 passenger trains between Ladybank and Mawcarse Junction were withdrawn – a goods service lingered until 1964. Passenger traffic on the Devon Valley line ended in 1964; all goods and mineral traffic ceased in 1973. The Alva branch lost its passenger service in 1954; complete closure came in 1965. Passenger traffic between Stirling and Dunfermline lasted until 1968; save for the Kincardine line, goods and mineral workings east of Alloa ended in 1979.

Abandonment of the Glen Farg line in 1970 emphasised the general railway retreat, with much of the road bed subsumed by the M90 motorway. Cowdenbeath-Kinross met a similar fate. However, there followed a restoration of sorts. The original Ladybank-Perth (Hilton Junction) line, singled east of Bridge of Earn in 1933, had lost its passenger service in 1955; but in 1975 Edinburgh-Perth trains via the Forth Bridge and Ladybank were re-instated and some Inverness trains have been so routed. Today's "Edinburgh and Fife" timetable approximates to an hourly pattern.

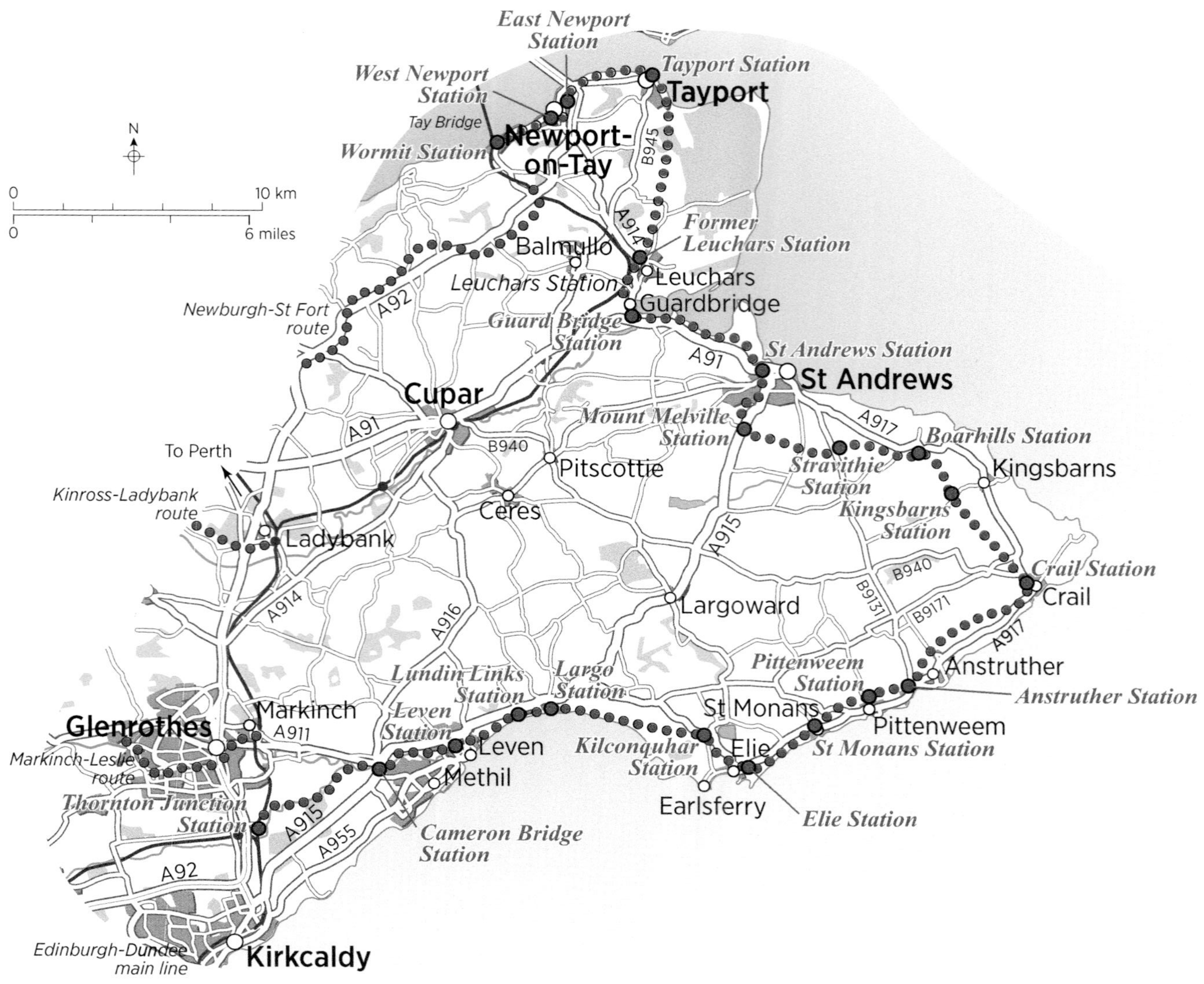
East Newport Station
West Newport Station
Tayport Station
Tayport
Tay Bridge
Newport-on-Tay
Wormit Station
N
0
10 km
0
6 miles
B945
A914
Former Leuchars Station
Balmullo
Leuchars Station
Leuchars
Guardbridge
Newburgh-St Fort route
A92
Guard Bridge Station
A91
St Andrews Station
St Andrews
Cupar
Mount Melville Station
A917
Boarhills Station
A91
B940
To Perth
Pitscottie
Stravithie Station
Kingsbarns
Kinross-Ladybank route
Ceres
Kingsbarns Station
Ladybank
A915
B940
Crail Station
Crail
B9131
Largoward
A914
B9171
A916
A917
Pittenweem Station
Anstruther
Lundin Links Station
Largo Station
Anstruther Station
St Monans
Leven Station
Markinch
Glenrothes
A911
Pittenweem
Leven
Kilconquhar Station
Elie
St Monans Station
Markinch-Leslie route
Methil
Thornton Junction Station
Earlsferry
Elie Station
A915
Cameron Bridge Station
A955
A92
Edinburgh-Dundee main line
Kirkcaldy

Cameron Bridge-Leven

The track bed from east of Cameron Bridge, plus a minor diversion, offers a pleasant 2½ walk to Leven, with an optional return by the River Leven path.

Map 59 covers the area.

From Windygates roundabout car park, map ref 350006, head briefly uphill on the A915, then descend left by a path to the track bed of the Lochty freight line. Few traces of that line are still mapped, but the tarmac path on the extant tree-lined embankment towards Kennoway gives a pretty out-and-back stroll.

On return, continue by tarmac path under the A915. The path turns right to join the Cameron Bridge-Leven line.

Now for another out-and-back stroll, following the easily-walked track bed west to Cameron Bridge station, clearly identified by the numerous platforms. On the left is the expanse of Cameron Bridge Distillery, the first to produce grain whisky. Continue on the

Cameron Bridge, in LNER condition *Lynn*

now overgrown out-of-use track which crosses then re-crosses the River Leven. On the second crossing two 10ft high padlocked gates bar the way.

Retrace steps to the merge point with the Lochty freight line. A lovely secluded walk, on the ballast path by the partly-hidden rusty rails, eventually leads to the 20-acre site of the Kingdom of Fife Railway Preservation Society at Kirkland Yard. The Society aims to create a heritage transport museum. From this point, the track bed has been completely obliterated by urbanisation, and the site of Leven station can only be identified by the Station Bar at the end of Station Road.

Retrace steps from Kirkland Yard or use the River Leven footpath.

Keep to the downstream side of the Yard on the old freight line to Methil power station. Where the Leven debouches into the Forth, pass under the single-span bridge opened in 1957 but still known locally as the 'Bawbee Brig', from a previous toll crossing where the charge was a Scottish halfpenny, or bawbee. A grassy area on the left leads to Levenmouth swimming pool/sports centre.

Cross the Bawbee Bridge and turn right, joining the Leven footpath on the south bank of the river. Cross the minor road bridge to the north bank, and then go left on a tarmac path close by the old creosote works. Cross again to the south bank, the site of a former steelworks, and continue upstream, passing through Kirkland Woods. Eventually cut down to the turbulent river at a weir and follow the path, which is on top of an old pipe line. Reach a footbridge and cross the Leven, then retrace steps to Windygates roundabout car park.

Leven-St Andrews

There is a curious but most pleasing dichotomy about the Leven-St Andrews railway walk. The southern half, along the more populous south coast, from Leven to Crail, passes by the numerous charming East Neuk seaside villages, whereas the north-westwards half, in staying well clear of Fife Ness, traverses a less populated agricultural area. Ideal for those middle class families who took summer quarters while the paterfamilias travelled daily to business, the prestige train was the Fife Coast Express, with both Edinburgh and Glasgow portions introduced in 1911. In its final version it survived until 1959. Though latterly a service of no great distinction, continuing to Dundee via Leuchars, it was allocated a coach-set from the LNER`s 1930s Coronation rolling stock.

Much of the track bed has been reclaimed for agricultural purposes. Many extant bridges can be identified from the map, but the surrounding ploughed-up track bed has transformed those bridges into reminders of what once was. A number of detours are thus required (mostly using parts of the Leven to Crail section of the Fife Coastal Path (FCP)... the 117-mile long distance footpath from Kincardine to Newburgh); yet, apart from the section west of Stravithie station, it is possible to walk all the way from Leven to St Andrews.

Map 59 covers the distance of nearly 30 miles.

Leven to Lundin Links station 2 miles

Though partially mapped, the track bed runs through three golf courses and it is incumbent on the walker to detour to the shore line and follow the FCP.

Leven, in LNER condition *Lynn*

An urban stroll leads to Scoonie at the eastern fringes of Leven. Cross the Scoonie Burn, pass by Scoonie Golf Club car park and follow the stonework-fringed path that winds by the edge of the golf course. Enter Silverburn Park, pass by cottages and turn right to follow the small burn by the Mile Dyke, which marks the boundary between Lundin Links and Leven Links golf courses, to reach the FCP.

On approaching the west side of Lundin Links, the FCP cuts inshore and crosses the golf course. Take note of the warning sign, "walkers may cross here but should be aware of the danger of golf balls from right and left". Turn right on a lane to reach Links Road and follow the northern fringes of the golf course. No evidence remains of the now grassy site of the single-platform Lundin Links station though one can guess the line of the track bed.

Lundin Links to Lower Largo 1mile

From Links Road turn right into Golf Road. The rail bridge over the road is no more but on the left is the embankment. Follow the embankment path and pass by Lundin Golf clubhouse and the large car park. Ignoring a FCP sign on the right, continue on a narrow path adjacent to the overgrown shallow cutting. Beyond the vanished overbridge at Drummochy Road, the track bed is overgrown then urbanised. Descend Drummochy Road but, before crossing the Boghall Burn, turn left up Burnbrae Terrace and then up steps to take a closer look at the impressive but unobtrusive three-arched railway viaduct. Retrace steps to reach the charming harbour at Lower Largo.

Lower Largo to Kilconquhar station 4 miles

Arguably no station-to-station walk has a more enchanting start than that on offer at Lower Largo. The historic fishing village, a designated conservation area, became a popular tourist centre after the arrival of the railway in 1856.That popularity, enhanced by the story of Alexander Selkirk, the inspiration for Daniel Defoe`s Robinson Crusoe, has outlived the railway. A Robinson Crusoe statue is displayed on a house in Main Street, the site of Selkirk`s birth.

Turn away from the harbour to again admire the grade B-listed viaduct, sadly out of bounds behind a high padlocked gate. Nevertheless, what a start and the walk too will not disappoint! Close to the harbour is the white-painted Railway Inn by the side of which the very narrow and attractive Station Wynd and ensuing steps lead to the site of the station and passing loop, now a landscaped area with a central gravel section. No buildings remain.

Thereafter the remarkably well-preserved track bed gives a charming stroll (albeit with one detour for a missing bridge), easing away from the eastern fringes of the village and so to open country. The track bed is briefly used by the FCP. Later the FCP stays by the curving shoreline of the wide Largo Bay, while the track bed continues in a straight line towards Kilconquhar. Pass by the dune grassland of Dumbarnie Links Wildlife Reserve where the track bed forms the northern boundary. Reach a gate and a row of re-aligned World War Two anti-tank concrete blocks that lead down to the shore and separates the reserve from the grazing land to the east.

Elie; note provision for fish and livestock traffic *Lynn*

Later the track bed is used by farm vehicles and there may be a herd of cattle. Pass a burn, on the left of which is a small pond, to reach a large barn and the site of Kilconquhar West, a passing loop opened in 1907. (Kilconquhar station had a curving platform which left insufficient space.) No trace remains of the isolated but impressive signal box.

With the A917 now closing in from the left, the track bed crosses the Cocklemill Burn and then the road to Shell Bay. No evidence of the level crossing gates remains. On the far side of the road, Kilconquhar station is easily identified by the lengthy single platform within an overgrown area. On the left is the stationmaster`s house, now a private residence.

The name of Kilconquhar is said to derive from the Gaelic *Cill Conchubair*, the church of Conquhar or Connacher. The name is pronounced locally as *Kinucher*, cf. Kinneuchar in 17th century records.

Kilconquhar to Elie station 1 mile

Beyond Kilconquhar station the track bed has been ploughed up, though the direction can be established by the road bridge over the line at St Ford. Beyond the bridge the track bed is detected again alongside the third hole of Elie`s nine-hole golf course. The railway then passed under the A917 on the western approach to Elie and continued through a cutting behind houses in Woodside Road before again passing under the A917 to Elie station.

The direct way to Elie necessitates a careful walk by the A917; however a more attractive detour is suggested. From Kilconquhar station walk with care along the A917 for 500 yards. Briefly turn left on the B941 then cross to a path which skirts the south side of Kilconquhar Loch. Head southerly on the track for Elie (not signposted), ignoring tracks in other directions. Continue past Elie House to emerge at the A917.

A slight rise in the road indicates the site of the overbridge. To the east is Baird Place, once the entrance to the station but now a housing estate. The station building suffered an early fate, set on fire by a member of staff in protest against the Beeching cuts. Continue west for a few yards to reach the former pedestrian access to the station. Behind the first house so reached is the identifiable boarded up arch of the overbridge and against the road wall is a pathway which led to a footbridge over the railway.

(For the more adventurous, a detour by the Elie Chain Walk, claimed to be Scotland's only via ferrata, may be considered. From Kilconquhar station follow the road towards Shell Bay then branch south to Kincraig Point. Amidst a jumble of rock pools and caves washed by the sea at high tide, the Chain Walk goes round the base of the Point in a series of carved steps, accompanied by vertical and horizontal chains. To maximise the time available, and for safety reasons, it is best to go when the tide is on the ebb. Check tide tables beforehand. Afterwards, take a sandy stroll east past Earlsferry Links.)

Elie to St Monans station 2½ miles

Beyond Elie station the track bed has been ploughed up. Walk down Wadeslea and turn left to reach the parking area at Ruby Bay, so called after the garnet or red gemstones washed up there. Later the gently undulating FCP goes parallel to the remnants of the track bed for a mile or so and passes by the fragmentary ruins of Ardross Castle. Once the track bed is lost to the fields, continue by the FCP. Further on are the more impressive and substantial ruins of Newark Castle where Alexander III (1241-1286) spent some of his childhood. Approach St Monans Church by a concrete path at the water`s edge (if the tide is in take a higher path) and cross the Inverie Burn.

(A stepped concrete path leads to Braehead and so to the harbour. St Monans, according to the locals, is a contender for the prettiest village in Scotland. I am inclined to agree. Return to the Inverie Burn.)

Head inland on the signed high-tide alternative path by the Inverie Burn to reach the mapped stone-arched bridge. To the west the track bed has been lost, but to the east the high grassy embankment leads by houses and a play area to the open site of the station south of the A917/Abercrombie road junction. All that remains to the right is one section of a brick-faced platform close by the entrance from Station Road. To the left is a small industrial area and a veterinary clinic.

St Monans to Pittenweem station 1½ miles

All traces of the overbridge at the A917 are gone and the entire track bed to Pittenweem is lost. The only reminder is the mid-point B942 overbridge. The A917 road-side path gives the most direct but less attractive route, so return to St Monans harbour and the FCP.

From Pittenweem, cross the A917 and head up Charles Street. Pass by the cemetery and Station Court (a small housing development) to a rise in the road by the speed de-restriction sign.....cue railway overbridge. All trace of the track bed to the west has been lost. To the east is the station cottage. Crossing the bridge gives access to the broad grassy site of the station.

Pittenweem to Anstruther station 1 mile

Beyond Pittenweem, a short stretch of track bed by a line of telegraph poles is easy to identify but the area is too overgrown. Thereafter the track bed is lost until the west side of Anstruther, so follow the A917 road-side path or the FCP.

Once in Anstruther, cross the A917 and head up Mitchell Place (west of the Dreel Burn), then follow a gravel track (signed circular walk to community woodland). Ignore a side track and continue on a gentle rise to reach the overbridge. To the east a new path by a landscaped play area follows the line of the track bed to the handsome broad twin-arched bridge over the Dreel Burn. Descend by steps for a closer look then cross the bridge, at the far end of which the track bed has been lost. Follow a pathway through the large industrial estate and the site of the station at the north-west side of the town, just west of the B9131. No trace of the station remains.

The local pronunciation of Anstruther, the largest community on the East Neuk, is "Enster". The open-all-year Scottish Fisheries Museum is well worth a visit. A warm and informative welcome is assured at the booking office/visitor-information hut on the quayside, from where cruises depart to the Isle of May.

Also Anstruther is famous for its fish and chips!

Anstruther to Crail station 4 miles

The railway took a northern curve well away from the A917, passing by Kilrenny Common, but the track bed has been lost to fields, including the short mapped section south of Thirdpart. Approaching Crail, the track bed, parallel to the A917, can be identified but is not walkable and is soon lost to urbanisation.

Similar developments east of Anstruther have covered the track bed, but follow the line of the railway past a car park and cross the B9131. Continue between the sports centre and primary school, on the right, and the new Waid Academy, on the left, to join a path which eventually follows a line of telegraph poles and is in fact the unmapped track bed. At a junction, map ref 567045, identified by an old metal post of railway origin, straight ahead the track bed is lost to fields. There are paths to the left and right; follow the latter, signed

Fife Core Path, to the A917 and so past Cellardyke`s small harbour, which, at the time of the herring boom, was one of the most important Scottish harbours. Looking out across the Forth, the Isle of May, a Nature Reserve, is easily seen. More distant is the Bass Rock.

The FCP then returns to the shore line with a caravan park on the left. The next section, the longest stretch of the FCP to be used, is very level with parts of the path rather too close at times to the sea. Later, pass the strange whorled and distorted sandstone cliff formation known as Caiplie Caves, reputably the site of early Christian worship. Continue along the cliff top overlooking Crail; a must-photo spot looking down to the small harbour and pretty houses. Crail station was situated at the northern end of the village on the west side of the A917. The roof of the station buildings can still be seen in the middle of what was for some time a market garden.

Crail to Kingsbarns station 2 miles

From Crail, just two miles south-west of Fife Ness, the most easterly part of the Kingdom of Fife, the railway headed northwards.

Take the A917 road-side path, parallel to the obvious but not walkable track bed to the B940 junction and the unmapped Sypsies overbridge. The track bed beyond has been ploughed up, apart from a section south of the mapped arched bridge on the B9171, map ref 600089. This again is not walkable; and the mapped path to Damside has been lost.

Continue with care by the A917 (the path soon comes to an end) then turn left on the B9171. Follow a lane adjacent to the track bed, signed Cycle route 1. Reach a horse rider-friendly gate and join the track bed, a grassy cyclable way used by farm vehicles. After a few more gates the track bed is pleasantly distant from the A917. Cross the infant Cambo Burn and then a Rights of Way path which goes left to Kippo and right to Kingsbarns.

A gradual descent leads to the road (initially signed Station Road from Kingsbarns) and the end of the cycleway. Turn right for a short distance on the road and then left by a lane, once the entrance to the site of Kingsbarns station. The lengthy platform edge lies amid an overgrown area and a few bricks remain of the signal box.

Kingsbarns lies one mile away, though the spire of Kingsbarns church is readily seen. The village name derives from the barns used for storing royal household grain when the King was in residence at Crail Castle or Falkland Palace.

Kingsbarns to Boarhills station 1½ miles

Heading onwards from Kingsbarns, hopes of a track bed stroll are quickly dashed. The modest embankment is too overgrown, although a mid-winter walk might be possible. For the worthwhile detour, head SW on the Kingsbarns road then turn right on the broad track, which can be driven. On the left are remains of concrete posts, the perimeter fence of the wartime Dunino airfield. It was occupied by 614 Squadron and 309 Squadron; the latter the only Polish Squadron to be formed in Scotland during World War Two. Finally sold in 1957, nothing remains of the grassy airfield although in the built up area, an eerie semi-wooded place, are the ruins of concrete buildings, some now reused for agricultural purposes. Head east to where the railway crossed the Kilduncan Burn. The track bed to the south can be walked for some 200 yards; to the north it is overgrown then lost to fields.

To reach Boarhills station there are two options: walk west to Upper Kenly, next north to Lower Kenly, then refer to the Boarhills-Stravithie section; or SE to the Kingsbarns road and so to Kingsbarns village. From there Stagecoach 95 bus service on the A917 (phone Traveline 0871-200-2233) gives a quick and easy transit to the entrance to Boarhills village.

Head back to the north side of the Kenly Water and follow the lane which passes under the Kenly viaduct (cue photo stop) to reach the mapped Park Mill, an impressive 18th century water mill. However, a parallel lane leads by houses to the site of Boarhills station. A modern house has replaced the station building. On my visit, when the house was up for sale, I was able to look round the well-tended garden area which extends east to the boarded-off far side of the viaduct; the track bed beyond is overgrown. In the other direction the track bed garden extends for a distance and then is similarly lost. In front of the house, the grassy platform is easy to identify. Another loading bank lies behind the house; the site of the railway yard.

Boarhills to Stravithie station 2½ miles

Head east with care on the A917 and cross the Kenly Water to reach a point opposite Falside Smiddy, from where a mapped path (a grassy track with fields on the left and the Water below to the right) leads to the railway embankment and a mapped underbridge. (The track bed to the right can be followed to the viaduct but the way is then barred and to the left is the overgrown track bed.)

Continue under the bridge on the lovely grassy way between fields to a T-junction. Turn right by the wooden sign, footpath, and descend to Lower Kenly. A tarmac track to the right leads to the bridge over the Kenly Water. Follow the minor road. The parallel track bed on the right has been subsumed by the field; but at the minor road from Bonnytown, map ref 556132, are abutments of the demolished railway bridge over the road.

Ahead the track bed, by the line of telegraph poles, is lost to fields and progress to Stravithie requires a lengthy and unsatisfactory detour, west past Bonnytown to the B9131. The latter road is very busy and car transport may be the safest option.

The site of Stravithie station (map ref 532134, a mile north of the hamlet) is clearly identified by the rise in the B9131 over the track bed and by the notice, Stravithie Country Guest House. The modest station building is no more. Stranded in the lovely grassy garden, close by the overbridge and a pond, is a railway carriage now converted into two luxury suites....resting on rails but with nowhere to go. Heading east, the lane on the north side of the overgrown cutting leads to open fields. The distant mapped bridge can be reached by following the field edge but that is scarcely a track bed walk.

Mount Melville, in LNER condition; single platform and goods yard *Lynn*

Stravithie to Mount Melville station 3½ miles

West of Stravithie, a lane parallel to the dark overgrown cutting leads to a bridge, beyond which the track bed is lost on the paltry ½ mile to the bridge south of Allanhill. A major detour, best tackled by car, is required. The track bed on either side of the Allanhill bridge has been lost to farming. Return to Allanhill, noted for its soft fruit, especially strawberries. Visit the farm shop at the right time of year!

Nevertheless, a delightful walk can still be found to reach Mount Melville. Head west on the broad lane to Lochend, also a right of way to Balrymonth and Lambieletham. This broad, charming, tree-lined way leads to open country and the first views of St Andrews. At the junction west of Scooniehill, turn south on the narrow path (Lambieletham 1.5km) to descend to a small building, the crossing keeper`s cottage. On the right is a signed footpath/right of way, Cairnsmill 1km, which follows the track bed.

Latterly adjacent to a burn, the track bed gives an excellent grassy walk. Pass by a very old and rusty turbine. The beautiful bush-lined path becomes a gently descending way, a hidden gem. Then a narrower path, slightly overgrown, winds through a wooded area alongside the small ravine of the burn to reach the A915.

The old A915, having curved west to the Lumbo road, then crossed the track bed at the mapped bridge, south of which was the site of Mount Melville station. Realignment of the A915 has obliterated all traces of the station....a sad end to what was a beautiful stretch of track bed.

Mount Melville to St Andrews station 2 miles

Beyond the Lumbo bridge, the track bed has been lost to pasture. Briefly walk by the grassy verge of the A915, then at a gate enter the southern end of Cairnsmill Holiday Park. Follow traces of the old road to the park office. Descend past the laundry from where a stepped path leads to a charming pond. Turn right on a grassy way by a lade to reach a bridge over the outflow. The path slants left (look back to the Lumbo bridge) then descends to the right to regain the track bed, now in a cutting. The track bed curves right but becomes overgrown and is then lost to urbanisation, so stay with the high path that meanders through the wooded Spinkie Den. Later the path descends to a burn to reach a grassy play area (No ball games!) and the first sight of houses on the southern side of St Andrews.

Ignoring the signed path, Lade Braes Walk, turn right on Bogwood Road which leads to the start of the lengthy Canongate; all new houses and no hint of history. Pass on the left first the entrance to the Botanic Garden (open all year round: cue for coffee stop) then the signed Viaduct Walk to regain the track bed.

The site of the railway bridge across the Canongate is evidenced by the abutment on the south side, where houses now occupy the embankment. On the north side, steps lead to a charming curving tarmac way, also used by cyclists. An embankment continues to the high but short three-arched stone viaduct over the Kinness Burn, best appreciated from below. Steps then lead into the town.

The way ahead, cutting through a series of car parks on levelled ground, still follows the line of the track bed. No longer are there overbridges at Argyll Street and Doubledykes Road.

Pass by Kinburn Tennis Club to reach a car park enclosed by high stone walls; the site of the constricted, island-platform station. The curving road to the right, Station Road, leads to the bus station.

Welcome to the Home of Golf, the Old Course and six other golf courses, yet for the moment all hidden to view. At the turn of the 19th and 20th centuries, railway companies became increasingly aware of the potential of 'golf traffic', offering special facilities, discounted fares and dedicated halts.

St Andrews-Wormit

Map 59 covers the 14-mile distance.

St Andrews to Guardbridge station 4 miles

From St Andrews station, keep to the right on the embankment tarmac path. The missing bridge over Pilmour Links, of which the abutments remain, necessitates a stepped descent. Cross the road and follow the west side of the embankment (at first with houses); it is later reduced to a small grassy way.

All of a sudden to the right is the Royal and Ancient Club House, the busy Old Course and the Swilken Bridge; a must-photo spot for golf aficionados. The famous small stone bridge spans the burn between the 1st and 18th fairways. In front is Old Station Road. Welcome back to the FCP which is then followed towards the Old Course Hotel.

The (relatively new) Old Course Hotel now occupies the site of the 1852 St Andrews Railway station which stood beyond the Old Course 17th hole and some distance from the town centre. The Royal and Ancient Golf Club had strongly opposed anything closer. Though restoration has been mooted, it seems unlikely that golfers would welcome the resurrection of a railway passing through the iconic Old Course.

With the arrival of the line from Leven, the old station was renamed St Andrews Links and the new one St Andrews (New).

It is extremely inconsiderate to cut through the golf courses by the plethora of paths and tracks. In any case most of the track bed has been lost and beyond the western extremity of the golf area, by the George Evans hide, it has been ploughed up. Accordingly, turn left as signed before the hotel and follow the FCP; hardly "coastal" on the stretch to Guardbridge.

Use the pavement on the old A91 which curves between the golf courses and a sports field to join its successor. There is then a fenced-off broad tarmac way between road and golf courses, part of Kingdom of Fife Cycleways/National Cycle Route 1, a busy and less than interesting, albeit safe, section of the FCP. Pass on the right a sandy track, signed to Eden Estuary Local Nature Reserve. (A grassy path in the Reserve leads to the track bed; to the east unwalkable and to the west subsumed by fields. Return to the road.)

When the separate FCP comes to an end, continue by the road-side pavement to the parking spot at the east end of Guardbridge. Although the line of the railway can be seen from the road, the track bed is lost to overgrowth, pasture and houses.

Follow what is the Old St Andrews Road, which passes by houses on the left and Guardbridge Inn on the right, to reach the Old Guard Bridge, completed in 1419 on the lucrative pilgrim trail to St Andrews. The six-arch stone bridge, spanning the tidal Eden estuary, was still carrying heavy traffic as late as 1938. Upstream is the new road bridge. Downstream are massive piers, all that remains of the railway bridge, best photographed with mandatory swans from the Inn garden. Cross the bridge and then the A919 to the site of Guardbridge station and coal yard, closed in 1965 and now completely subsumed by houses.

Guardbridge to Leuchars Junction 1 mile

From Guardbridge a narrow pedestrian/cycle path, roughly on the line of the railway, leads through the small housing estate. The path curves by the back of other houses to regain a short stretch of track bed, but, beyond a grassy area, heading NNW to a mapped bridge just short of Leuchars Junction, it is lost to the fields. However, the walk is not over.

Continue by the A919 pavement to the old Inner Bridge and turn left up River Terrace by the charming, reed-covered Inner Burn. Just before Motray Park bowling club, curve right on a grassy path by a derelict building. On the right is the disused Guardbridge Reservoir, on my day almost bone dry. A few yards further, behind a brick wall, are the remnants of the rail bridge that spanned the Inner Burn, its piers almost submerged in the reed-covered water. To the left is a padlocked footbridge, dangerous and to be avoided.

Beyond the piers is an inaccessible embankment. Nothing remains of the trailing junction with the Edinburgh-Dundee main line. A restored branch, say protagonists, should be laid out for through running from Edinburgh to St Andrews. Hidden to view, Leuchars station lies further north-east beyond the Milton bridge.

Leuchars Junction to Tayport station 5 miles

A remnant of what was the original Burntisland-Tayport rail line, the track bed goes in a straight line across an extensive, level area (note the lack of contours) first drained in the 1700s. Once past the western side of the extensive Tentsmuir forest, it curves past Morton Lochs to reach Tayport harbour.

(Tentsmuir allegedly was so named following the shipwreck of a Danish fleet in the 1780s when some of the sailors settled there, living in tents on the moor. However, even in 1791 the parish minister was pooh-poohing stories about shipwrecked Danes, explaining that it was the inhabitants of the inland part of the parish who coined the name after seeing the tents of shepherds on the muir.)

From Leuchars roundabout, follow the gravel track between the Commercial Arms Hotel and new houses to reach gates and so to the track bed. Dogs should be kept on lead or under close control at all times. Roughish and grassy at first, the way improves on reaching the southern end of a narrow wooded strip. Indicative of the flat cultivated area, the track bed runs parallel to a deep drainage ditch, at one point spanned by a bridge, the sides of which are lined with short sections of rail.

Note the warning of frequent shooting in this area. Reach the level crossing over the road to Rhynd and Kinshaldy. On the far side, the grassy way, latterly tree-fringed, leads in little

Leuchars (Old): on original Burntisland-Tayport main line *Lynn*

more than ½ mile to a barrier and the end of the mapped track bed. There is no option but to retrace steps, thus dividing the Leuchars--Tayport walk into two halves.

The next access to the track bed uses the very poorly signed entrance to Morton Lochs from the B945 at map ref 453230. Once over the road, take care to slant left and so to the National Nature Reserve. Three lochs (North, South and West) were created by diverting the Lead Burn, originally known as Ninewells Burn, and flooding the dune slacks (hollows) and surrounding wet lands. Pass by South Loch and turn right for the car park, built on the track bed, beside an attractive stone-arched overbridge. The track bed can be followed south but only for some 150 yards; it is then overgrown, at least in summer.

At South Loch is a signposted footpath to Tayport, which goes by North Loch. However, from the car park continue north on the track bed, a pink-coloured wide path. When that path curves to the left, continue on the grassy track bed until, as mapped, it is lost to the fields ahead. Follow a narrow curving path by a line of fence posts on the left to join the signposted footpath, in an area of ancient sand dunes. Continue north by a gate through a wooded area, pass by Garpit and through Scotscraig golf course, giving consideration to golfers.

Exit the golf course and turn left on Shanwell Road for an urban stroll to Tayport harbour. (The track bed is lost to a small industrial estate and extensive housing developments.)

All traces of the harbour-side station have been erased. The only evidence of the site of the station is a wall mounted plaque "Erected on the site of Tayport railway station, commemorating the visit in 1877 by US President Ulysses S Grant who then embarked from Tayport harbour to view the construction of the first Tay Bridge". Round the corner is Tay Street also signed President Grant Place. It was a wild, wet day according to press accounts and the notoriously taciturn Grant was also obliged to walk out from Dundee esplanade over the completed spans. "It`s a very long bridge", was allegedly his only comment.

Tayport, Newport East and West stations to Wormit station 4 miles

From Tayport harbour, go over what was once the level crossing. A sign indicates Tay Road Bridge 2, Newport 3 and Wormit 4 miles, a distance covered by the Fife Coastal Path (FCP).

Go along Inn Street by the water`s edge, follow the FCP between houses and eventually gain the track bed tarmac path; a straight and gently rising, secluded wooded way between coast and road. With views opening out over the Tay Road Bridge, the track bed comes alongside the B946 but is then lost to fields. Follow the FCP road-side tarmac way.

At Newport, the railway took an elevated course by the eastern fringes of the town. However, the track bed has been mostly lost to the Tay Road Bridge approaches, the A92 realignment and urbanisation. To reach the sites of both Newport stations, it is better to descend under the Tay Road Bridge on the coast road and then seek directions, or to be well organised with a street map.

Second Tay Bridge; piers of Bouch's narrower, top-heavy viaduct; typically Scottish signal with lattice post and finial *Lynn*

Surrounded by a white picket fence, the stone building at odds with new houses, the former Newport East station house, is 13 Station Brae on the corner with East Station Place, reached from the B995.

Situated on a short lane where the top of Shepherd`s Brae turns right on Bridge Street, the Newport West station house is in an architectural style to be seen again on the Newburgh and North Fife Railway. The site of the station, now a grassy landscaped play area, has a large information board.

From the station, it is a joy to be back on the track bed to Wormit; an excellent stretch to finish what may have been for some the long walk from Leven!

At a road bridge, the way ahead may appear to be blocked but it weaves left past private garages and continues as Station Path, parallel to the track bed. Traverse a sports centre car park and at the far end slightly descend under a road bridge and on to the now signed Newport and Wormit Nature Trail; a partly secluded way heading for the Tay Rail Bridge.

After a cutting, a missing bridge over Flass Road is easily circumvented. Though a house blocks the track bed, a path leads to a road. Turn left and pass between the high abutments of the vanished bridge. Continue along Birkhill Avenue where the parallel track bed is lost to gardens, and so to a brick wall overlooking the gated-off overbridge at the coast road.

Descend by Hillpark Road to the coast road. Houses cover the site of Wormit station on the far side of the overbridge. Further east, turn down Bay Road towards the high brick approach arches that carry the main line to the Tay Rail Bridge. From there steps lead to the station of which the only evidence is a small section of brick wall.

In 1955 the locomotive of a Tayport-Dundee train overturned at Wormit tunnel mouth. The train was a school excursion returning from Tayport. That the ex-LMS locomotive,

running tender-first, was `foreign` to this ex-LNER (NBR) line, may have been a factor. Excessive speed, alcohol and an adult and child `unauthorised` in the cab were all noted. The eight non-corridor coaches carried 540 passengers, mainly children. After the tender derailed on the tight curve through the tunnel, tender and engine, and the two leading coaches, overturned within Wormit station. The fireman and the cab `guests` were killed. The driver and forty passengers were injured, more or less seriously.

To mark the walkers' achievement, perhaps in finishing the long walk from Leven, it would be appropriate to celebrate in The View Restaurant, Naughton Road; once a pub called the Tay Bridge Halt.

Having made his reputation, not least in respect of the Forth and Tay train ferries, civil engineer Thomas Bouch established his own firm. During the 1860s his forte was the 'cheap railway', though the results were not always happy, and the Leuchars-St Andrews branch is a case in point. The challenge of the Forth and Tay water-breaks continued to fascinate him.

The collapse of the first Tay Bridge (28th December 1879), the biggest civic engineering disaster in British history, destroyed the reputation of Bouch who had been knighted for its design and construction. He died 18 months later.

Dundee's unofficial Laureate, William Topaz McGonagall, immortalised the disaster, which hurled the Sunday evening mail train, railwaymen and passengers, into the Tay:

"The Storm Fiend did loudly bray,
Because ninety lives had been taken away,
On the last Sabbath day of 1879,
Which will be remember'd for a very long time."

Recent scholarship suggests that the gale, though a contributory factor, was not exceptionally severe. The consensus is that derailment of the train proved the last straw for a vulnerable, already over-stressed structure. This partially-but only partially- vindicates Bouch`s own interpretation of the disaster.

Bouch`s bridge, and its successor, were entirely a North British commitment, though a subsidiary company was created to further the original scheme. However, the Forth Bridge Railway Company was a joint enterprise, and the three English partners of the North British contributed a large proportion of the capital.

In the aftermath of the Tay Bridge disaster, work on Bouch`s Forth Bridge ceased. The new Fowler-and-Baker cantilever design insured heavily against the vagaries of wind and weather. In recent years, Bouch`s Tay Bridge design, generally condemned after 1879, has been more objectively studied, while the back-to-back suspension bridges which he planned for the Queensferry passage have been more favourably reassessed.

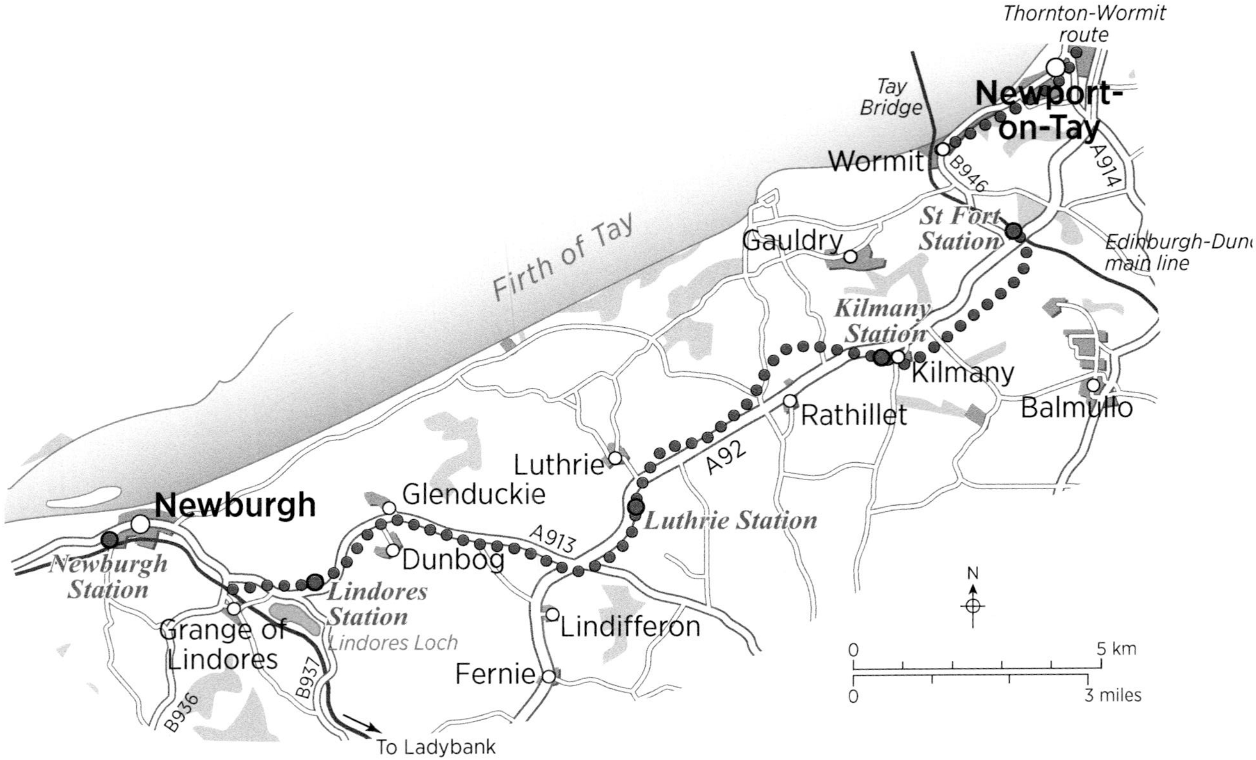
Thornton-Wormit route
Newport-on-Tay
Tay Bridge
Wormit
B946
A914
St Fort Station
Edinburgh-Dun main line
Gauldry
Firth of Tay
Kilmany Station
Kilmany
Balmullo
Rathillet
A92
Luthrie
Luthrie Station
Glenduckie
Newburgh
A913
Dunbog
Newburgh Station
Lindores Station
Lindores Loch
Grange of Lindores
Lindifferon
Fernie
B937
B936
To Ladybank
N
0
5 km
0
3 miles

Luthrie: passing loop removed; concrete platform walls speak of late construction *Lynn*

Glenburnie Junction (Den of Lindores) -St Fort Junction

That most of the relatively short-lived Newburgh & North Fife railway has been recovered for agriculture is a great disappointment for the walker. The vanished overbridges (many abutments remain), along the A913 and the A92 on side roads and farm lanes, have fragmented the track bed into short or overgrown lengths. An overview by car is perhaps the best choice. Nevertheless some aspects make exploration on foot worthwhile, though only the portion of the route which survives as the Kilmany Railway Line Heritage Path offers a worthwhile stroll.

Map 59 covers the area.

Glenburnie Junction (Den of Lindores), on the Ladybank-Perth line, has been completely obliterated. Nearby is must-visit tranquil Lindores Loch, a Site of Special Scientific Interest and once an important source of waterpower for mills in the Lindores valley.

At the north-east end of Lindores village, a lane leads to a bridge over the track bed. Follow the west-side grassy way to a modest tree-studded embankment and charming reedy ponds. Lindores station house, still so named, is the white building by the A913 and east of the ponds. A wooden fence at the back of this private dwelling is supported by short lengths of old rails. No evidence remains of the bridge to the east over the A913.

South of Luthrie village (a common approach to Norman's Law) and east of the lane to Easter Kinsleith, is Luthrie station house; called The Firs, though Station House is painted

on the garage door. Other homes now occupy the whole site. Only a small section of platform edge in one landscaped garden remains. Just east of the station, a track leads to the partly overgrown, best-done-in-winter embankment which can be followed to the abutments on the A92. The high embankment beyond the A92 gives a short out-and-back stroll when approached from the lane to the NE.

Further east, through a potato growing area, the track bed has been ploughed up as far as Rathillet House. Cross the Motray Water and turn right, parallel to the overgrown cutting, on the road to Stirton, and so to the next underbridge, finest of all the structures on this line. The ornate brick-lined skewed arch is a thing of beauty. Turn left just before the bridge to gain the embankment and cross the bridge.

Initially overgrown, a winding path on the high wooded embankment leads to a metal gate and an open grassy way, frequented by cattle, overlooking the red pantile houses at Stirton Mill. Beyond a wooden barrier lies a veritable jungle. Save perhaps in winter conditions, a return to the bridge is prudent.

Continue by the A92 to the minor road, signed Gauldry and Balmerino, and the high abutments of yet another bridge. On the right is a Fife Core Path sign. This is one mile of the

Kilmany Railway Line Heritage Path, likely to need regular strimming. The path on the tree-lined small embankment continues under the A92 to a hidden, deep and wooded cutting. Pass under an old bridge to reach the open site of Kilmany station, now occupied by three houses and landscaped gardens. Opposite is the now extended and modernised Station House, the mansard roof just visible behind a high hedge. Jim Clark, the former world champion racing driver, was born in the village. Look out for the memorial sculpture unveiled by Sir Jackie Stewart in 1997. (See also Duns station on the St Boswells-Reston Walk.)

The track bed continues to the south side of the village and can be regained, beyond a missing bridge, for a final half-mile. When it ends abruptly, return to the A92, from where, a little further to the north-east, can be seen the graceful double-arched bridge over the railway at Wester Kinnear.

Turn off on the B946 to the sign pointing right, Moffat &Williamson, bus and coach hirers. This business occupies the old railway yard, the site of St Fort (North Junction). Follow the road under the Edinburgh-Dundee main line for a closer look at St Fort station house, now a private residence.

Kinross-Milnathort-Mawcarse Junction-Glen Farg

The Dunfermline--Glen Farg--Perth line brought the North British Company advantages in passenger traffic between Edinburgh and the Highlands. There was also heavy coal traffic from Fife to the North, and the Glen Farg gradients were a proving ground for the latest locomotives of the early 20th century.

Mawcarse Junction remains *Crawford*

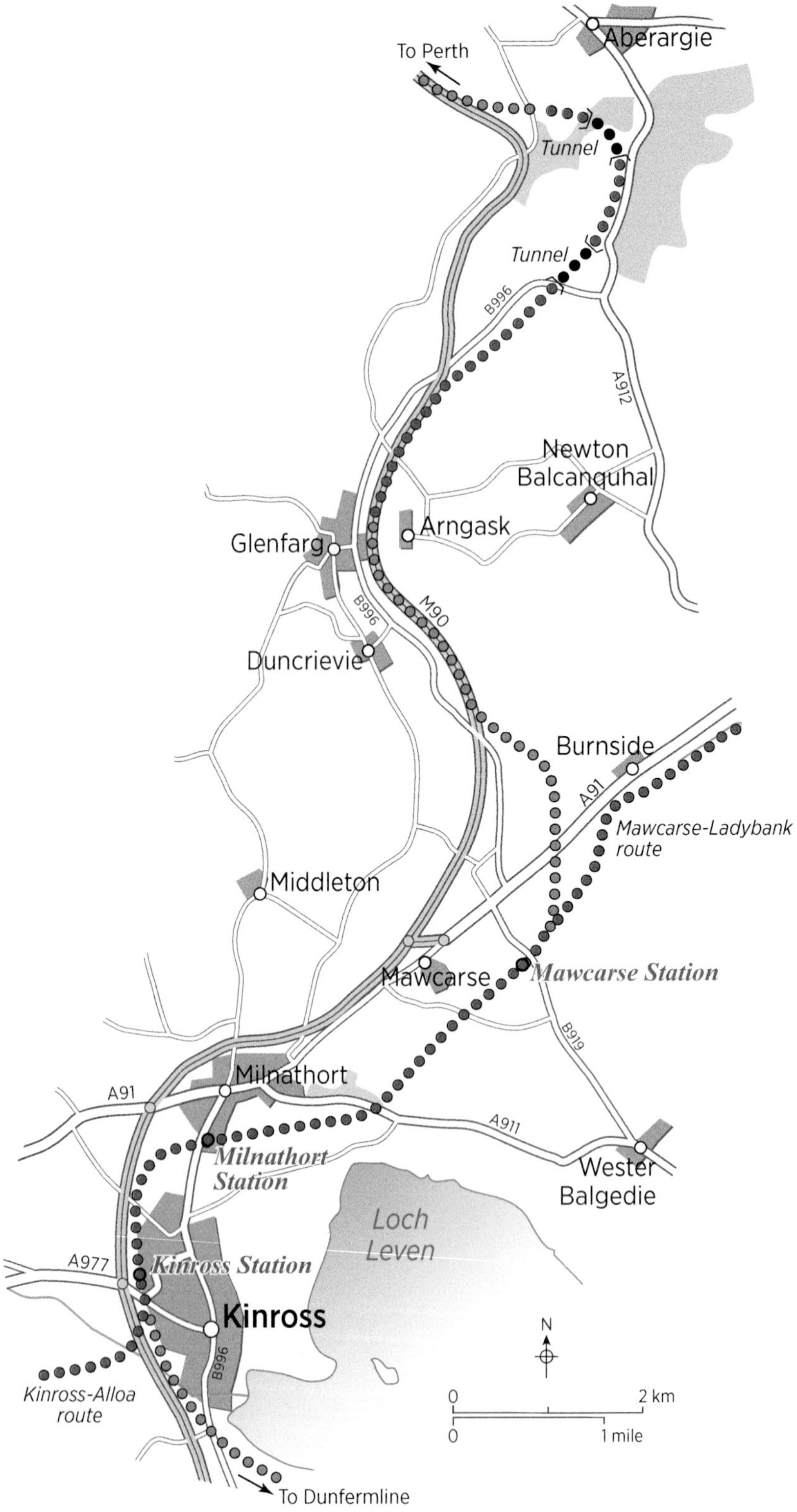

Aberargie
To Perth
Tunnel
Tunnel
B996
A912
Newton
Balcanquhal
Arngask
Glenfarg
B996
M90
Duncrievie
Burnside
A91
Mawcarse-Ladybank
route
Middleton
Mawcarse
Mawcarse Station
B919
Milnathort
A91
A911
Milnathort
Station
Wester
Balgedie
Loch
Leven
A977
Kinross Station
Kinross
B996
N
Kinross-Alloa
route
0
2 km
0
1 mile
To Dunfermline

Lost to urban development and road modernisation, little is left of the line and such former stations as Kelty, Blairadam and Kinross. Furthermore, much of what is mapped as dismantled railway is not walkable or, if walkable, is not of satisfactory length, as on the SW side of Loch Leven adjacent to the Loch Leven Heritage Trail.

Even so, there is still a short section round Kinross; a longer stretch from Milnathort towards Mawcarse Junction; and then the hidden gem of the Glen Farg tunnels.

Map 58 is required.

Kinross ½ mile

Kinross railway station, on the west side of the town, close to the sharp turn of the B996, map ref 112028, is no more. However, the nearby track bed, round the north-west side of the town, has become a ½ mile tarmac path, well-used by local residents. Beyond a minor road, the non-walkable track bed curves towards the distant but visible bridge at Milnathort station.

Kinross Junction in NBR condition (posters for two other pre-Grouping companies) *Lynn*

Milnathort to Mawcarse Junction 2 miles

The readily identifiable site of Milnathort station, beside the Auld Mart, is at the southern end of the town where the B996 crosses high above the old line. On the north side of the bridge, a lane slants down past Railway Cottage; but for access to the track bed, go to the south side of the bridge, from where Auld Mart Road leads by modern houses to a small park. With the North Queich Burn on the right (this flows into Loch Leven) and Milnathort Golf Club's nine-hole course on the left, the excellent track bed path takes a direct easterly course.

After ¾ mile, at a signposted junction, the track bed is joined by National Cycle Network, Route 1, part of the Kingdom Cycle route, Burgher Bridge to Mawcarse.

Thereafter the tarmac path/cycleway follows the modest embankment between fields, passes by a path on the right to Burleigh Sands and the Loch Leven Heritage Trail, and then, making a small diversion to avoid farm buildings, reaches the A911. Cross the road to regain the track bed.

Reach a sign-posted lane on the left, and the end of the track bed, lost in the fields ahead. Go down the lane to Mawcarse Farm and turn right by road to an overbridge, map ref 146055, from where, beyond more fields, can be seen a distant high bridge over the B919, the site of Mawcarse Junction, map ref 153060. This is reached by road via Lomondville.

Amid the wide grassy site, only a small overgrown platform remains as testimony to a railway past. Beyond the bridge, the grassy expanse forks between trees (northward for Glen Farg and eastward for Ladybank), recalling the vanished junction.

Glen Farg 2½ miles

If coming from the south, follow the B996 through the village of Glenfarg, under the M9 and into wooded Glen Farg for ½ mile. Park on the left hand side just before the road crosses a burn, map ref 150126. The line can be followed for a short distance back towards the M9 to a spoil heap left from the motorway construction. A few typical fence posts survive.

Return to the start point and then go deeper into Glen Farg, where the double-track railway was uncompromisingly, yet not too obtrusively, imposed on the landscape. Cross a few awkward fences to continue through a cutting. The line then leads to a deeper cutting, crosses high above the B996 and so to the first tunnel, generously proportioned (a reminder that the Glen Farg cut-off was a relatively late addition to the railway network).

On my day the immediate entrance was being used on either side as a rubbish dump, including one car, but thereafter the way was clear. Without a torch (I know, very stupid) it was a slow shuffling but easy walk in the mostly dry but very dark tunnel. Even with a torch, it might be better to go on a bright day. On leaving the tunnel, the immediate way ahead was somewhat muddy and overgrown. However, with the road on the right below, the next gently graded section, fringed by silver birches, is the prettiest part of the walk.

The second tunnel was almost debris free. Back into daylight and with open fields on either side, a considerable embankment then leads to a viaduct over a deep gully and so to a minor road. With the bridge over the road now removed, this is "the end of the line". A short distance away is the M9 into which the former railway disappears.

Mawcarse Junction-Ladybank

Much of the 11½ mile distance, which roughly follows the infant River Eden (already encountered far downstream at Guardbridge), has been lost, leaving fragmentary sections of variable quality. Nevertheless, there are still hidden stretches of great charm.

Maps 58 and 59 cover the area.

Mawcarse Junction to Gateside station 3 miles

From the wide grassy expanse of Mawcarse Junction (refer to Kinross—Glen Farg Walk) heading NE for Ladybank, the line of the railway is easily determined, but the track bed is lost to fields and pasture and, beyond that, to a fenced-off plantation.

From the Junction, head south on the B919 and turn left on the minor road past Butterwell towards Burnside. At a tarmac lane, map ref 163073, descend past the National Grid, Milnathort gas installation point, to regain the track bed. Well worth doing despite the noise from the A91, the ½ mile out-and-back trip is the first of those hidden stretches. Beginning by a wooded strip, the well-worn track bed is used to access fields on either side. At the edge of the plantation, pause and retrace steps.

Continue by road to the bridge over the track bed and on to Burnside. Follow the new Burnside path on the right hand side of the A91. (For an easy continuation, follow the A91 to the Bannaty Mill lane and descend to regain the track bed.)

The path turns right and descends to the River Eden, here just a burn. The path is intended to continue to the Lomond Hills but shortage of funds, at the time of writing, has stopped progress. The remnants of the old bridge (a few sections of rails) and, on the far side, a very overgrown traverse to a minor embankment, will not deter the adventurous walker, though this is best done in winter. Perseverance is rewarded in the shape of a superb track bed walk.

Beyond a gap in the embankment, the track bed, used for farm access, gives another secluded stretch, one mile downstream to Bannaty Mill. It continues, latterly in grassy form, to a missing bridge over the Eden; avoided by following a grassy local path, which slants up to Station Road

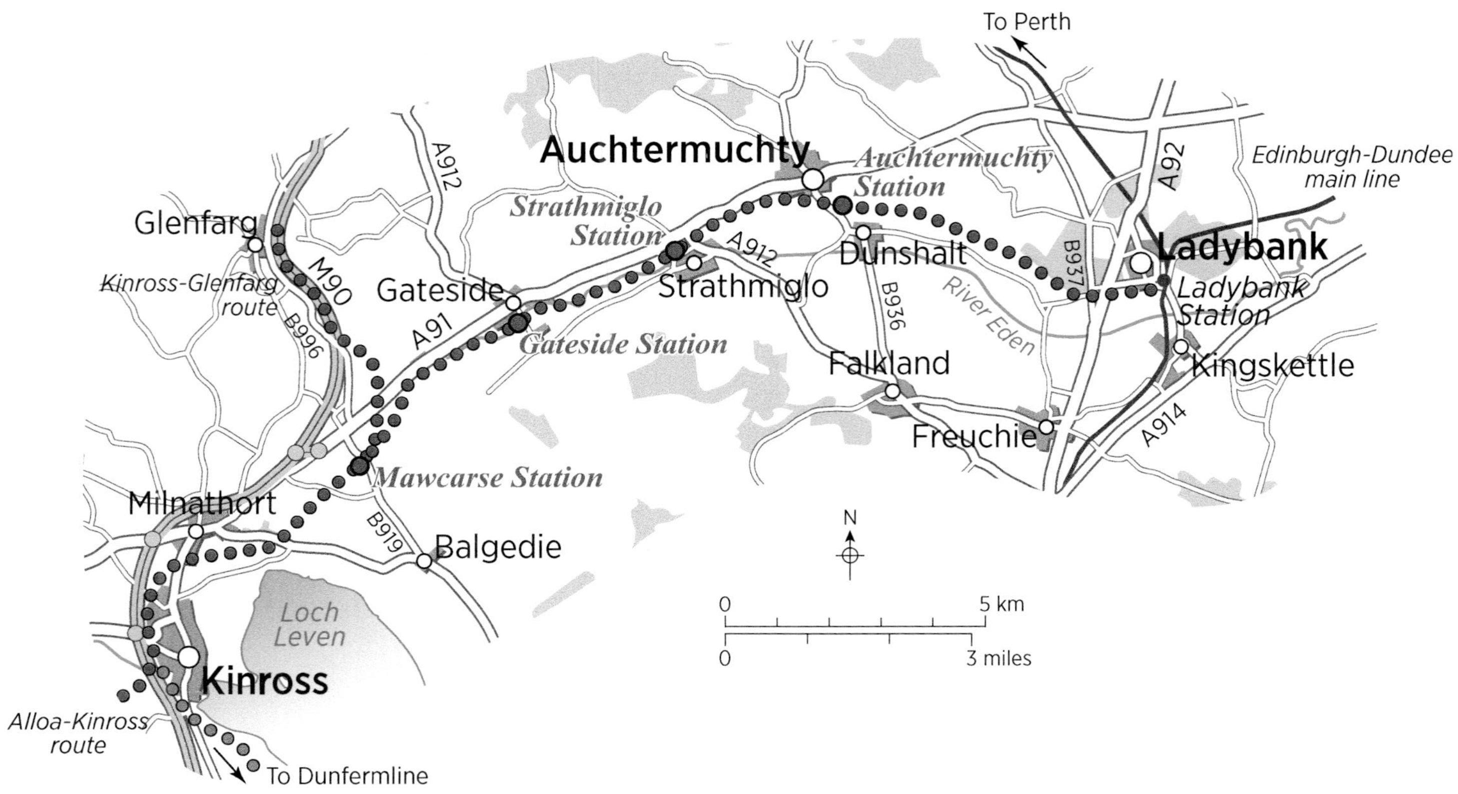

To Perth
Auchtermuchty
Auchtermuchty Station
Edinburgh-Dundee main line
A92
A912
Glenfarg
Strathmiglo Station
Ladybank
Dunshalt
B937
Kinross-Glenfarg route
M90
Gateside
A912
Strathmiglo
Ladybank Station
B996
A91
B936
River Eden
Gateside Station
Falkland
Kingskettle
A914
Freuchie
Mawcarse Station
Milnathort
B919
Balgedie
N
Loch Leven
0
5 km
0
3 miles
Kinross
Alloa-Kinross route
To Dunfermline

Descend to the obvious road bridge straddling the track bed and the Eden. On the right is Gateside station house (the name is above the door), which has acquired an incongruous extension. The only other evidence of the former railway is a section of platform edge in the garden.

Gateside to Strathmiglo station 2 miles

A stepped path leads to the track bed and yet another tree-fringed hidden stretch by the Eden. After ½ mile turn left to Gateside Mill, nowadays a derelict area. Wooden bobbins and shuttles were once manufactured here, largely for the Indian jute and hessian industries. Timber, once delivered by rail, latterly came by road. The rails of the spur can still be seen.

(The next ¾ mile downstream to Corston Mill, by cutting and embankment, is intermittingly walkable; gorse and a missing bridge are challenging. A grassy embankment comes to an end and ahead the track bed is ploughed up. From a stone-arched bridge, a lane passes by the abandoned buildings at Corston Mill to reach the A91.)

From Gateside Mill, head to the A91and follow the left-hand pavement to Strathmiglo. Cross the A91 to the immediate overbridge. (Back towards Gateside, the line of the railway, adjacent to the old A81, ends beyond a broad grassy area.) On the left side of the bridge a winding path leads through Strathmiglo Community Woodland; the site of the station,

of which no evidence remains. Station Road, just beyond the Strath Tavern, leads to Old Station House, hidden behind a high hedge and a fence made of railway sleepers. (The grassy expanse in front is divided between a paddock and the school play area.)

Strathmiglo to Auchtermuchty station 2 miles

The Fife Core Path in the Community Woodland, winding by many a silver birch, roughly follows the line of the track bed adjacent to the A91. (Still worth doing, the path extends beyond the A912 but, after meeting the unwalkable track bed, the path loops back.) Carefully cross the A91 and follow the left-hand pavement all the way to Auchtermuchty. (From a high overbridge, map ref 222109, it is readily apparent that the track bed is too rough, and then lost beyond the Barroway Burn.)

Stroll through Auchtermuchty, musing perhaps about the town's famous sons.... identical twins Craig and Charlie Reid (the Proclaimers, think of "Sunshine on Leith") and accordionist. Jimmy Shand, who played traditional Scottish dance music. Turn right on the B936 and along Station Road to the eastern outskirts of the town. The industrial area on the right is the site of the Fife Ironworks, established in the 1870s and once Scotland's largest. The foundry ceased production in the early 1990s.

Continue to where the hump-backed road crosses the railway line. The overbridge is now filled in. In 1950 a railway worker was killed when the road bridge collapsed under the excessive weight of a 26-ton tractor trailer carrying a 30 ton caterpillar-track digger. East of the bridge, the track bed can be followed for a short distance, latterly between unfenced fields.

North of the bridge is the entrance to a Business Park. At the far end is the extensive site of the Sterling Furniture Auchtermuchty Distribution Centre where the offices have incorporated the station building; the structure saved but not its dignity. I was privileged to be given a private look around but left with little impression of the old station building. Few artefacts remain. Outside is a truncated platform edge with the original steps. The literally circumscribed station frontage, now with a floor above, can be viewed from the road bridge.

Auchtermuchty to Ladybank Junction 4½ miles

It is likewise disappointing to discover that over the fertile flat ground between Auchtermuchty and Ladybank Junction almost all of the track bed has been ploughed up. A station-to-station walk is out of the question, nevertheless the Rossie Drain, dug in 1806 to drain Rossie Loch and recover useful farm land from Rossie Loch, is worthy of a visit.

Follow the B936 to Dunshalt, then east on the minor road towards Ladybank. At a lane, map ref 277098, a grassy path to Kilwhiss leads between remarkably flat fields, crosses the lost line of the track bed and rises to a stone arched bridge over the Rossie Drain.

Auchtermuchty, in LNER condition; prominent poster advertising 1938 Empire Exhibition *Lynn*

Cross the bridge, restored in 2004 with assistance of the Heritage Lottery Fund, and turn right on the northern side of the Drain. Best left for winter days to minimise the undergrowth, it is a slow 400-yards traverse to reach the western end of Heatherhall Wood, wide coniferous woodland with many a path for walkers.

Follow the path alongside the surprisingly deep Drain by the SW edge of the wood and parallel to the line of the lost track bed. Eventually reach a parking area, map ref 292094 (just west of the B937), situated by an open grassy way between trees where the line once ran. If the 400-yards traverse above is not appealing, simply do an out-and-back stroll from the car park.

From the car park there is no trace of the track bed to Heatherinch level crossing, where the Kinross line, as first built, converged with the then Burntisland-Tayport main line. This constituted a second junction at Ladybank, where the original Perth line diverged. However, to close on a happier note, do visit Ladybank station, reckoned to be the oldest unaltered station in Scotland still in operation.

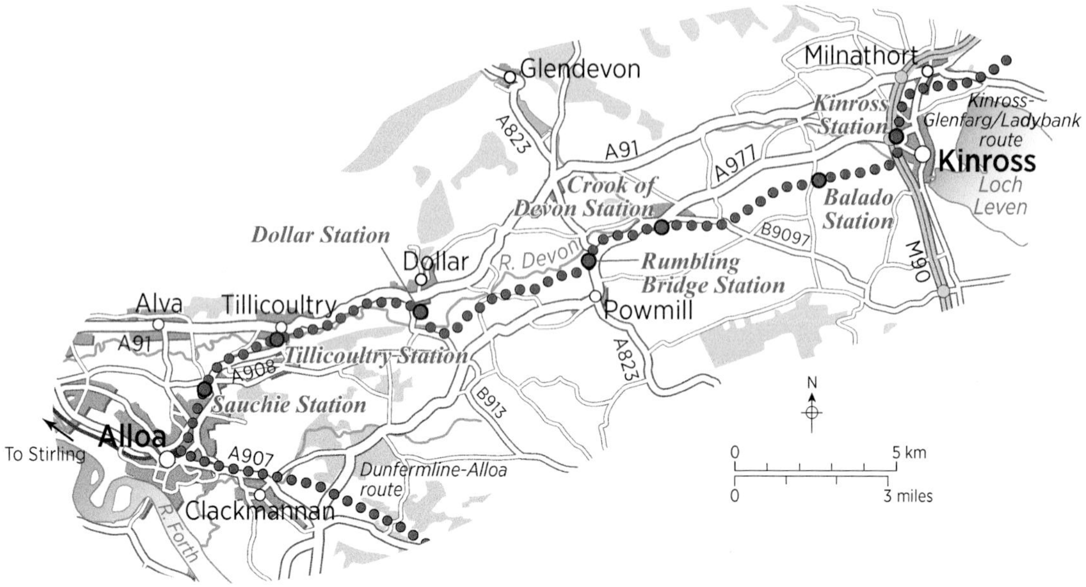

Kinross-Alloa

Despite being just 17 miles in length, this is one of the most varied railway lines and the 14-mile walkable stretches and detours have much to commend them.

In contrast to the easy levels around Kinross, following the River Devon meant heavy construction work. The most challenging part, the 4½-mile section between Rumbling Bridge and Dollar, involved inter alia construction of a 900-yard embankment leading to a viaduct over the Devon and a second viaduct over the Gairney Burn.

Map 58 is required.

Kinross to Balado Station 1¾ miles

Kinross station was situated on the west side of the town, close to the sharp turn of the B996, map ref 112028. Most of the track bed over the flat ground towards Crook of Devon has been lost. However, the mid section around the site of Balado station (map ref 085013) offers pleasant linear walks. At South Balado, now so styled, the station, on the east side of the level-crossing over the minor road between the A977 and the B9097, is identifiable from the remains of one platform wall.

Heading west, the track bed gives access to a house and then to fields. After one mile, with a clump of trees on the right, reach a barrier across the trackbed. Beyond, the general line of the grassy way can be identified but the effort is scarcely worthwhile and in any case the vicious barbed wire would deter all but the most intrepid.

Heading east, the track bed, much used by tractors and walkers, gives easy walking for ¾ mile between a new plantation and open fields. The track bed beyond is lost. To extend the walk, turn left on a grassy path between fields. Keep to the path skirting a big house on the right, then by tarmac road to the old A977 by the outskirts of Balado. Turn left on the road/path and left again as signed Coldrain 2 miles.

Crook of Devon station to Rumbling Bridge station 2 miles

Crook of Devon lies where the Devon, after flowing eastwards from the Ochils, takes a sharp turn westwards. Two miles downstream is the deep and narrow Rumbling Bridge Gorge whence the Devon meanders its way to the Forth near Alloa.

The track bed between the villages has been subsumed by farming and both stations are lost to housing developments, though their sites can still be identified. Nevertheless, it would be a pity not to walk station-to-station, using the signposted path by the lovely banks of the Devon.

Rumbling Bridge in NBR condition; horse-and-cart in modest goods yard *Lynn*

Crook of Devon village hall is situated at the east end by the War Memorial. A few yards to the south is the site of the station. (The rough grassy embankment to the east may look inviting, but soon becomes overgrown with saplings and is then subsumed by fields.)

Head west, pass new houses on the right, and follow a narrow path by the edge of the field to regain the track bed. Pass Waulkmill Park. Ahead the track bed is lost, blocked by more housing. A path swings right, then left down a lane to meet the former railway. Turn right on the path to Powmill and Rumbling Bridge, and follow the lane towards Harelaw. Later turn right as signed, Rumbling Bridge 1¼ miles and cross the A977 where all evidence of the overbridge has been lost.

A stepped path on the far side leads to the Devon. Follow the curving riverside path to the Rumbling Bridge gorge and on to the A823 and Rumbling Bridge, where the Devon is spanned by two bridges, built one above the other. (The rumble comes from the river, not the double bridge!) The lower bridge, 80ft above the river, was built in 1713; the upper bridge, 40ft higher, in 1816. It is worthwhile to descend into the gorge by the gate at the north end of the bridge.

At the Old Station sign, go down what was the entrance to the station which is now identified only by the remains of a platform on the left. The way ahead is blocked by houses.

Rumbling Bridge to the Devon Viaduct 4 miles

An option is to head west to Vicar`s Bridge by the road on the north bank of the Devon, then east up the minor road which meets the track bed (map ref 988979); but the intrepid walker will not wish to bypass the secluded delights of the viaduct at Gairney Burn and a deep narrow cutting.

From Rumbling Bridge, turn right on the minor road to Nether Dalkeith farm and Gean Cottage. Pause at the bridge that overlooks the former line, now part of a garden. To the west, the immediate track bed is lost though the next grassy embankment is in sight; reached by a track turning left from Gean Cottage. The high embankment at first offers rough walking, then easier going, to reach a lesser cutting, now blocked, where the Craigton farm track intersects. Regain the track bed, pass a rough tree-fringed area and so reach the sapling-covered Gairney viaduct. In picturesque surroundings, this impressive five-span structure, 110ft high, straddles the deep, V-shaped Gairney Glen.

Follow the rim of another cutting, waterlogged then blocked at a farm lane. Short detours are later required, as two small bridges are gone and an embankment has been partly washed away. Then for the second delight: a metal gate marks the entrance to the mossy but dry cutting; a secret way hidden from the nearside minor road.

Go under a 3-arched sloping bridge, the way to Arndean, then under an ornate estate bridge by the mandatory rhododendrons. A short, rough stretch then leads to where a bridge once spanned the minor road from Vicar's Bridge, map ref 988979.

A track leads back to the former railway and the start of a superb stroll; enjoyment enhanced by the difficulties of the previous stretch, plus the views to the Devon and Vicar`s Bridge. A small cutting, slightly damp, with brick retaining walls, leads to a bridge which carries the second minor road from Vicar`s Bridge to Mains of Blairingone. (Blairingone was an old mining area.) To the west the massive embankment on the north bank of the Devon can now be identified.

The track bed enters a wood and the immediate cutting is deep and overgrown, with wet stretches. However, persist to a tree-fringed embankment, high above the river. Beyond a brick wall are the piers of the Devon viaduct. An easy descent leads to the river. Though the Devon, when low, might be forded, prudence dictates a return to the end of the wood from where a grassy way, past a new plantation, leads to the B913; then followed to Dollar.

Dollar station, via Tillicoultry and Sauchie to Alloa East Junction 6 miles

From Park Place on the east side of the Dollar Burn, go past Strathdevon Primary School, Dollar Health Centre and playing fields, to reach the track bed. Initially turn left, then right,

to traverse the massive grassy embankment leading to the Devon, and which gives a close look at the piers (one now with a couple of trees) of the once high and imposing, multi-arched 390 ft viaduct. Return and head west for Dollar station, where one long platform survives.

Squeezed between the A91 and the Devon, and overlooked by the Ochils, a tarmac path (the Devon Way and cycle route 767) utilises much of the track bed and gives an easy stroll to Alloa.

Pass under the B913. The first stretch is through a tree-fringed cutting, hidden from the A91. At a signpost, back the way, Dollar ¾ mile, the line crosses the minor road from Lower Sheardale. At the far side is another sign, Alloa 5 miles, Tillicoultry 2 miles. Thereafter the close-by A91 is both seen and heard, while on the left is the meandering Devon. With an almost imperceptible descent from Dollar, look out on reaching Tillicoultry for an old white marker which gives the gradient as 1 in 436. Cross the A908 to the site of the station, now a grassy recreation area, and on to a path junction.

Take the path curving to the left (Alloa 3 miles, cycleway 767). This crosses the Devon by an arched footbridge, and then goes parallel to the A908 to regain the track bed. However, the railway walker will curve right for a few yards (Menstrie 4 miles, Alva 2¾ miles, cycleway 768) to reach a high embankment leading to the brick-built piers; all that remains of the

Dollar after closure – 'farewell' special (BR twin-car DMU) *Lynn*

demolished Glenfoot viaduct where the railway re-crossed the Devon. Follow the river bank to reach the 1889 Glenfoot Bridge and head south on the road-side pavement. Turn left on a stepped path and regain the track bed, a tree-fringed straight and pleasant way between the A908 and the Glenfoot road.

After crossing the road, the cycleway path closely follows the line of the track bed and continues through a silver birch-fringed area. Beyond two stone-arched bridges (the second one carries the B9140) is the site of Sauchie station.

BR coal train at Tillicoultry *Furnevel*

A long straight stretch leads to the north end of Alloa. Cross a broad grassy area to reach a roundabout where the railway bridge is no more. Head right and follow the path/767cycleway (¾ miles to go) to regain the line of the track bed and reach the expansive site of Alloa East Junction. Before reaching the high 3-arched bridge, follow the cycleway path on the left to cross the footbridge spanning the restored Stirling-Alloa line and the new Alloa station.

Dunfermline-Alloa

Maps 58 and 65 are required.

The double-track Stirling-Alloa-Dunfermline-Thornton line carried heavy coal traffic, and Dunfermline shed, like Thornton shed, had a large allocation of locomotives in the days of steam.

There were five intervening stations on the 11-mile stretch between Dunfermline and Alloa---from east to west Oakley, East Grange, Bogside, Forest Mill and Clackmannan Road. All traces of Dunfermline Upper station have been obliterated, subsumed by a large car park, houses and a retail park. A short section of track bed, a local signed pathway, heads east, passing between a cemetery and Dunfermline Athletic football ground, to reach Townhill Junction, where the line from Inverkeithing trailed in. (This survives as part of today`s Fife Circle, via Dunfermline and Kirkcaldy and vice versa.) Townhill Junction is now styled Dunfermline Queen Margaret and Dunfermline Lower styled Dunfermline Town.

Heading ¾ mile west of Dunfermline Upper station, all trace of the track bed has been lost until reaching William Street. The track bed from this point to the site of Clackmannan Road station,1½ miles east of Alloa station, has been converted into a tarmac way used by walkers, runners and horse-riders (note the frequent mounting blocks) and offers glimpses of Fife`s industrial heritage in passing by former mining villages. The most popular part of the tarmac way is the initial five-mile section from William Street car park, where a useful notice board makes mention of Dunfermline`s past as "resting place of Scotland`s kings and their queens".

A path bypasses a row of houses occupying the line of the former railway, then regains the track bed which rises gently to the line`s modest 100m summit which offers more open views over farm land.

3½ miles from Dunfermline is the site of Oakley station, just west of where the track bed crosses the road (Station Road!). Only overgrown remnants identify the platforms and the exchange sidings that served Comrie Colliery. The Colliery branch itself has been overtaken by urbanisation and overgrowth. On the south side a path leads to Inzievar Woodland walks. Oakley ironworks, once employing hundreds of men, ceased production in 1869.

Bogside signalbox and siding, in BR condition *Lynn*

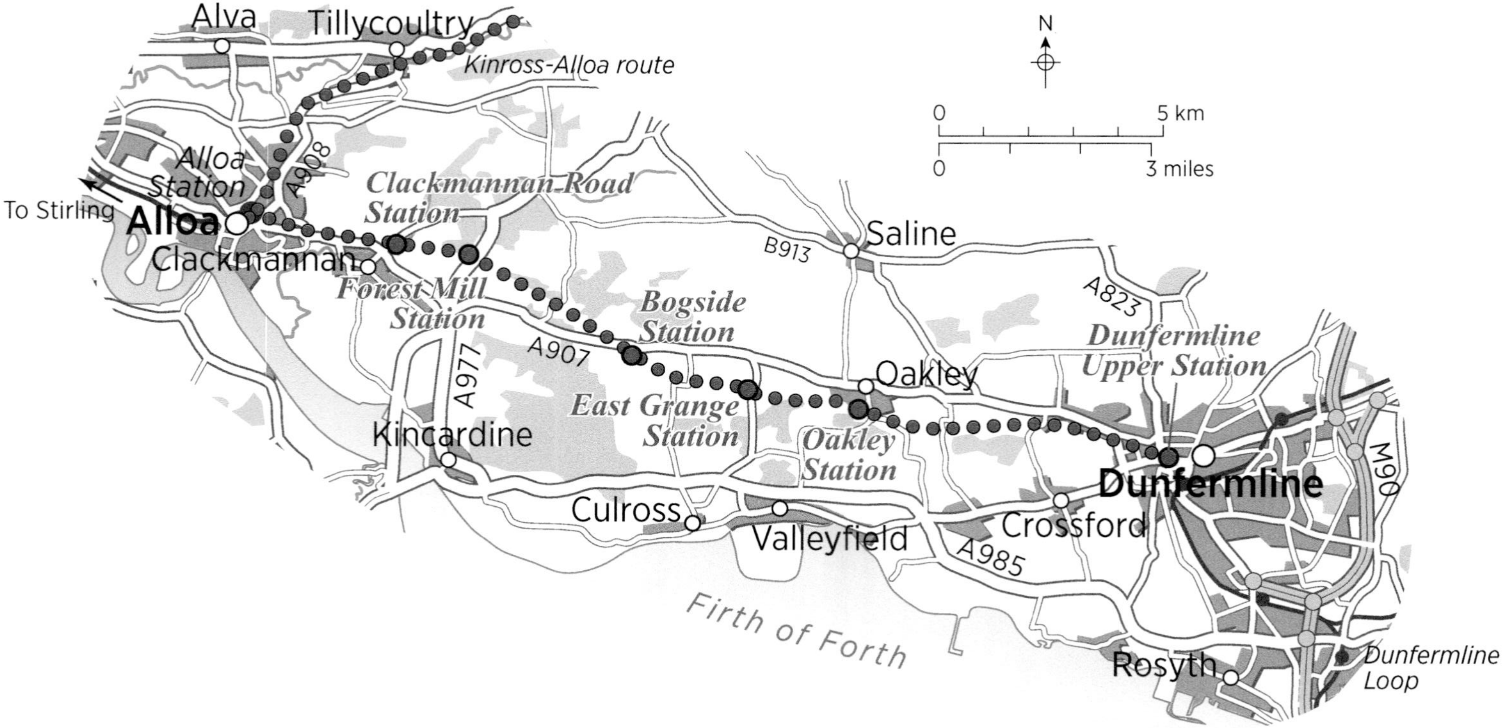
Alva
Tillycoultry
Kinross-Alloa route
N
0
5 km
0
3 miles
Alloa Station
To Stirling
Alloa
Clackmannan
A908
Clackmannan Road Station
Forest Mill Station
Saline
B913
Bogside Station
A907
A977
A823
Dunfermline Upper Station
Oakley
East Grange Station
Oakley Station
Kincardine
Culross
Valleyfield
Crossford
A985
Dunfermline
M90
Firth of Forth
Rosyth
Dunfermline Loop

It is then a gentle descent by a series of long rock cuttings through a surprisingly well-wooded area above the deep gorge of the Comrie Burn. Cross Comrie Dean Viaduct and continue to where the track bed crosses the B9037—the eastern side accessed by path from the village of Blairhall. East Grange station was situated on the west side but no trace remains.

The track bed slants NW on a high embankment and crosses the Bluther Burn by a substantial bridge. A secluded tree-fringed way leads to the site of Bogside station, just south of the A907. On the left, nowadays in a ruinous state and partially hidden by bushes, is the wooden waiting room for Alloa-bound passengers. Further on is Bogside signal box, all

that remains of the station and indeed is the sole railway building on the entire line to have survived.

A 2½ mile straight stretch through a mixture of open and wooded country passes by the mapped parking area at Slack to enter Clackmannanshire. A high viaduct takes the track bed over the A977; on the west side is the site of Forest Mill station, evidenced only by the remains of the left-hand platform.

Another straight section of one mile leads past a tall brick-built chimney, remnant of the Cherrybank Brickworks, which closed in the 1950s; the southern side is well worth examining.

Then cross the high Black Devon viaduct and follow the embankment to the site of Clackmannan Road station (of which there is no evidence) just east of the B910.....and the end of the tarmac way. Although the bridge over the B910 is no more, the far-side

Kincardine Junction (Clackmannan): modern signal and abandoned trackbed, with crouching hare *Furnevel*

The Dunfermline-Alloa line via Kincardine, not shown on the area map (p.160), remains in being; its future is uncertain.

embankment can be regained for the final ½ mile; a pleasing stroll on a winding grassy path as far as the track to Hillend Farm. Descend to the old road parallel to the track bed from where the Kincardine line bridge over the A907 is well seen.

It is appropriate that the walk ends after crossing the Black Devon. The area encapsulates the industrialisation of Fife, the collapse of the mining industry and the subsequent reclamation of industrial wasteland. The hamlet of Forest Mill, two miles to the NW, straddles the Black Devon: here water was diverted by weir and lade to Gartmorn Dam Reservoir (1713), which fed the water-wheel pumps draining the collieries in Alloa. The dry lade can be followed by a heritage path to Gartmorn Dam, now a Country Park & Nature Reserve and a Site of Special Scientific Interest. Coal extraction and the removal of clay caused so much subsidence on the last two miles of the Black Devon's flow to the Forth that the river simply disappeared. However the mixture of fresh and salt water has been utilised to establish the Black Devon Wetlands. The site is leased to the RSPB.

Cambus-Alva

Subject to a diversion by the A91 pavement, the full length of the track bed from Cambus to Alva can be walked. The four miles can be extended by detours to the Tullibody Old Bridge and Menstrie Castle.

Map 58 is required.

Cambus, the southern suburb of Tullibody, is bounded to the west by the River Devon. The road across the Devon is adjacent to the boarded-off Cambus Iron Bridge; a rare Scottish example of a single span prefabricated lattice girder bridge. 68 feet in length, it was constructed in the early 19th century to link with the west bank distillery, which was founded in 1806 but closed in 1993. (The far bank is now an extensive area of bonded warehouses.) Still in evidence on the deck of the Cambus bridge are the guide rails, gauge 4ft, for the horse-drawn wagons. ("Scotch gauge", before Stephenson`s standard prevailed, varied between 4ft and 4ft 6".)

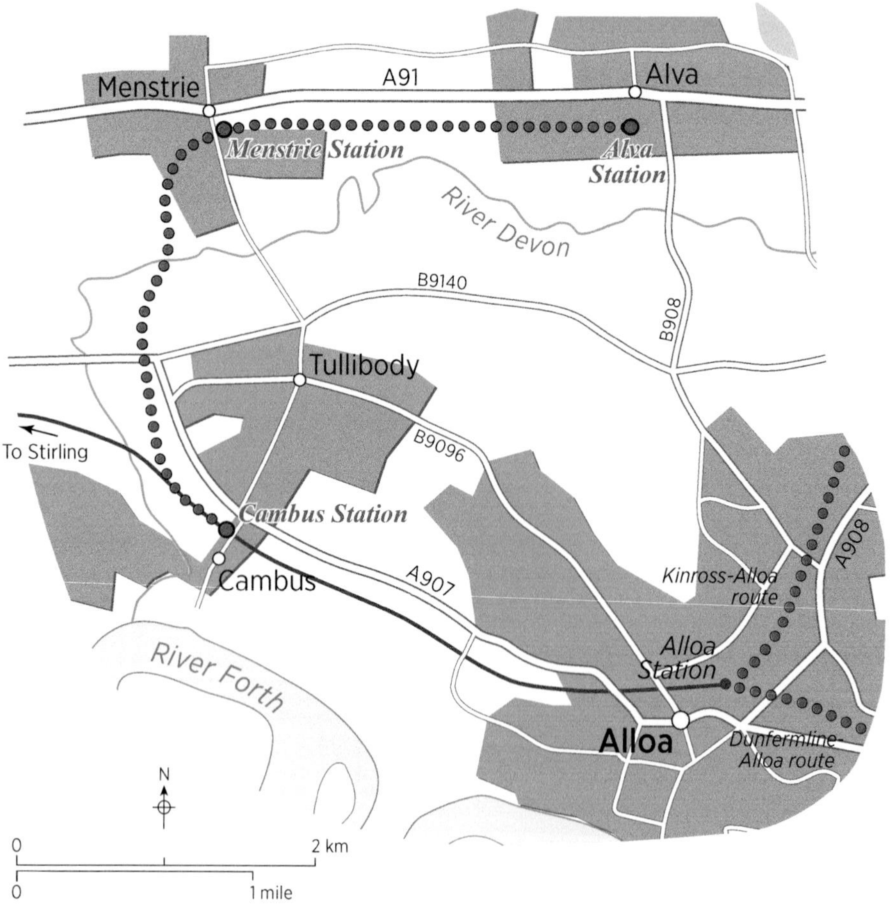

Menstrie in LNER condition: the nameboard reads 'Menstrie & Glenochil' *Lynn*

The broad tarmac path by the east bank of the Devon is liable to flooding and the river at first is not at its most attractive. The path goes under the restored Stirling-Alloa railway to join the branch track bed, then curves right again to pass by the impressive ruins of the 17th century New Mills Crossing Doocot.

The straight track bed passes under the graceful arched stone bridge carrying the old A907, then the unattractive concrete bridge carrying its successor. (From this point, make a

brief detour westwards along the old A907 to the Tullibody Old Bridge, an early 16th century toll crossing on the Devon. Consisting of two main spans and three flood arches, it is 442 ft long and varies in width from 11ft to 30 ft. Redundant from 1915, it was restored for use by walkers and cyclists in 2003.)

Return to the track bed and head north, Menstrie 1¼ miles. With the Ochils as a backdrop, cross the Devon (a lovely secluded spot not seen from any road) after which the track bed curves east on its attractive approach to the minor road on the east side of Menstrie. The track bed passes under the road to the former site of Menstrie & Glenochil station of which practically nothing remains.

The now grassy track bed can still be followed for a short distance but soon is lost within the extensive grounds of the once independent Glenochil Distillery, now part of Diageo. Retrace steps to the station site, and then walk into Menstrie to have a look at Menstrie Castle, birthplace of Sir William Alexander, 1st Earl of Stirling, who established a colony in Nova Scotia in the 17th century. More castellated house than true castle, it stands bizarrely amid modern houses.

Head east by the A91 pavement for almost one mile to a lane, map ref 868970, and then regain the track bed. This can be followed back west for a short distance to a burn just before the Diageo perimeter fence. Return and continue east on the track bed/path by the south side of Alva to reach a grassy play area—the site of the old station at the foot of Queen Street and only identifiable by a building signed Station House.

Markinch-Leslie

The Leslie and Auchmuty feeder lines had the local paper mills by the River Leven as their main source of income. Indeed, until recently paper-making was a major employer in Glenrothes New Town. However, the closure of the Tullis Russell plant (itself a merger of the former Auchmuty and Rothes mills) in 2015, left just one paper mill still in operation, the Sapphire Mill in Leslie.

Although parts of the track bed have been lost to road developments and housing, the 5-mile route can still be broadly detected from a study of the map. Embankments were needed to approach two viaducts across the gorge of the Leven. What is left of the track bed is now a broad tarmac way suitable for walkers and cyclists (part of National Cycleway 766) and generally well-signposted; map 59 is not really needed. Much of the way is more secluded than the map suggests. The section through Glenrothes is known as the Boblingen Way, a reference to Glenrothes` twin-town in Baden-Wurttemberg, Germany.

Turn left on leaving Markinch station and follow the old track bed/tarmac way. On the right is the former Haig`s whisky bottling plant, now a business centre. After ¼ mile a sign Balbirnie Viaduct path confirms the railway`s ancestry. Then reach the former Auchmuty Junction from where the branch line descended towards the Leven and Auchmuty.

Passing the signs, Leslie Railway Company 1861 and Leslie 4½ m, Markinch ½ m, the Leslie line gradually rises to Balbirnie Bridge Viaduct. On the next section the track bed has

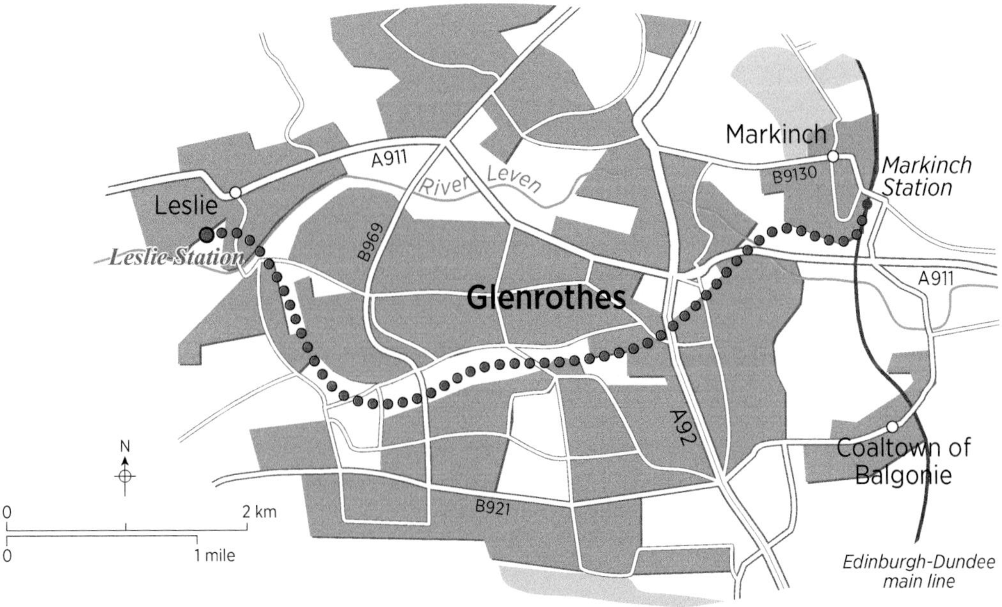

all but disappeared; but signposts help and the general direction is SW. Turn left by Alburn Crescent and eventually pass under the A92. On reaching the Lomond Centre on the right, cross a road and turn left to regain the signposted track bed, Leslie 3 miles. The tarmac way curves through Warout Wood, on the right of which is Glenrothes Juniors football stadium. Beyond Lomond Business Park, on a steady descent, the Lomond Hills come into view.

Cross a busy road and so to the curving 14-arched, 214 yards Leslie Viaduct. (The Earl of Rothes had apparently objected to the railway crossing his land, hence the sweeping curve of the line to the deep gorge of the Leven and the subsequent need to build such a high viaduct.) A listed structure, it was designed by Bouch, of first Tay Bridge notoriety. The original cast iron railings and low flat-coped parapets were removed during its conversion to carry the Boblingen Way.

Beyond the viaduct, the track bed continues to the site of Leslie station. Only a few parts of an old wall remain, where a siding came in from the nearby paper mill.

Given that the railway succumbed to road competition, it is ironic that an easy return, rather than retracing steps, is to take a bus to Glenrothes centre bus station, then on again to Markinch.

South-Central Highlands

Strathyre's heron fountain (see Section *Callander to Strathyre*) *Crawford*

North through Stirling the obvious rail route to Dundee and Aberdeen, or into the Highlands, ran by Strathallan and Strathearn, avoiding the barriers presented by the Firth of Forth and the Firth of Tay. It was direct for Glasgow traffic and, in due course, for Anglo-Scottish trains via Carlisle. The Scottish Central Railway (Greenhill-Stirling-Perth) was completed in 1848 and would absorb the Dundee & Perth Railway in 1863. North from Perth, the Scottish Midland Junction Railway and the Aberdeen Railway – which in 1856 combined to form the Scottish North Eastern – ran by Strathmore and Stonehaven to Aberdeen. Greenhill was approximately the mid-point of the Edinburgh & Glasgow Railway and at Greenhill the Scottish Central joined end-on with the Caledonian Railway giving connection with Carlisle via Carstairs. (The Edinburgh & Glasgow's Polmont-Larbert line subsequently provided a better Edinburgh-Stirling link.) Hampered by break-of journey at the inevitable ferries across Forth and Tay, the Edinburgh, Perth & Dundee Railway (1848) was not a dangerous competitor.

Early proposals for a railway into the Highlands via Crieff came to nothing. In 1856 the Scottish Central Company provided their own branch from

Gleneagles (called Crieff Junction until 1912). The Perth, Almond Valley & Methven Railway was built that same year, as a Scottish North Eastern client. The Strathtay & Breadalbane Railway (Perth-Aberfeldy) of 1845-6 had been lost – it was overtaken by the Inverness & Perth Junction Railway, a too-ambitious 100-mile Spey-to-Tay line which Parliament rejected; but the modest Dunkeld Railway (1856), diverging from the Scottish North Eastern at Stanley, subsequently succeeded. Though the Inverness companies eventually achieved a tenuous connection with the South via Aberdeen (1858), they were determined to achieve a more direct route across the Grampians. On a second attempt (and a little altered), the Inverness & Perth Junction was authorised (1861) and speedily completed (1863). The Ballinluig-Aberfeldy branch was soon added (1865). By agreement with the Scottish North Eastern, the Inverness & Perth Junction subsumed the Dunkeld Railway and gained entry to Perth from Stanley.

The amalgamations of the mid-1860s saw the Inverness companies united into the Highland Railway and brought the Caledonian Company and the North British Company to Perth where Caledonian, Highland and North British now shared the General Station. In respect of traffic exchanged, the North British would remain disadvantaged until 1890, when the Forth Bridge was completed. (In the interval Anglo-Scottish traffic via Edinburgh reached Perth by Larbert and Stirling, by arrangement with the Caledonian.) Moreover, the Highland Company could scarcely be impartial – their Stanley-Perth access now rested on the goodwill of the Caledonian as successor to the Scottish North Eastern.

The Crieff & Methven Junction Railway (1864) had been a Scottish North Eastern protégé, arousing Scottish Central suspicions that their neighbour nursed wider ambitions. With both companies absorbed by the Caledonian, these frictions disappeared and Crieff became a local railway centre of some importance. Extension westwards, by two Caledonian subsidiaries, to Balquidder Junction on the Callander & Oban Railway (below) came late in the day. The Crieff & Comrie Railway opened in 1903, the Lochearnhead, St Fillans & Comrie Railway in 1905. A likewise late addition, under the Act of 1896, was the Bankfoot Light Railway, diverging at Strathord from the Caledonian (Scottish North Eastern) main line north of Perth. The Highland Company more than once considered, but never pursued an extension of their Aberfeldy branch to Kenmore on Loch Tay.

The Dunblane, Doune & Callander Railway, first proposed in the 1840s, had been revived with Scottish Central encouragement, opening in 1858. And it was with the assurance of Scottish Central backing that the promoters of the Callander & Oban Railway, authorised in 1865, chose a cross-country route to the west coast. (The alternative route via Loch Lomond, under Edinburgh & Glasgow patronage had been rejected after much debate). When the Scottish Central's obligations passed to the Caledonian Company, the latter proved less than enthusiastic – which partly explains the slow progress of construction, halted first at Glenoglehead (for Killin) in 1870 and then at Tyndrum in 1874. With the Caledonian at last thoroughly committed, the line was completed to Oban

Comrie, in CR condition
R.Maxwell, courtesy D.Ferguson

in 1880. The Killin Railway (1886), locally promoted and locally owned, remained distinct. It was, however, in practical terms a Callander & Oban branch, worked, like the main line, by the Caledonian. (Strictly speaking, the Killin Company and the Callander & Oban retained their independence until the LMS absorbed them both.) Killin Junction was simply an exchange station, without road access. A modest tourist traffic passed between the Caledonian and Highland systems by the Loch Tay steamers. The tourist trade of the Trossachs found a railhead at Callander – with the Strathendrick & Aberfoyle Railway (a North British subsidiary) half-partner, half-competitor from 1882. (A Callander & Oban branch by Loch Venachar to Loch Katrine was costed but nothing resulted.)

Perth-Methven-Crieff passenger services ended in 1951, goods traffic in 1967. (The original Methven dead-end closed in 1937.) Comrie-St Fillans-Balquhidder was another 1951 closure. Gleneagles-Crieff-Comrie succumbed in 1964, after a brief experiment with a cheap-to-operate railbus. The Aberfeldy branch, the Dunblane-Callander line and the eastern half of the Callander & Oban, together with the Killin "branch", were all closed in 1965, as the Beeching Plan gathered pace. (Westward from Crianlarich, the Callander & Oban was reduced, in effect, to a branch of the West Highland Railway.) Passenger services on the Bankfoot line were long gone (1931) but it survived until 1964, for the annual potato traffic. Callander-Crianlarich and Killin Junction-Killin suffered a more abrupt end than was scheduled, following a major rock-fall in Glen Ogle. The Killin locomotive and rolling stock were retrieved via Crianlarich and the West Highland line.

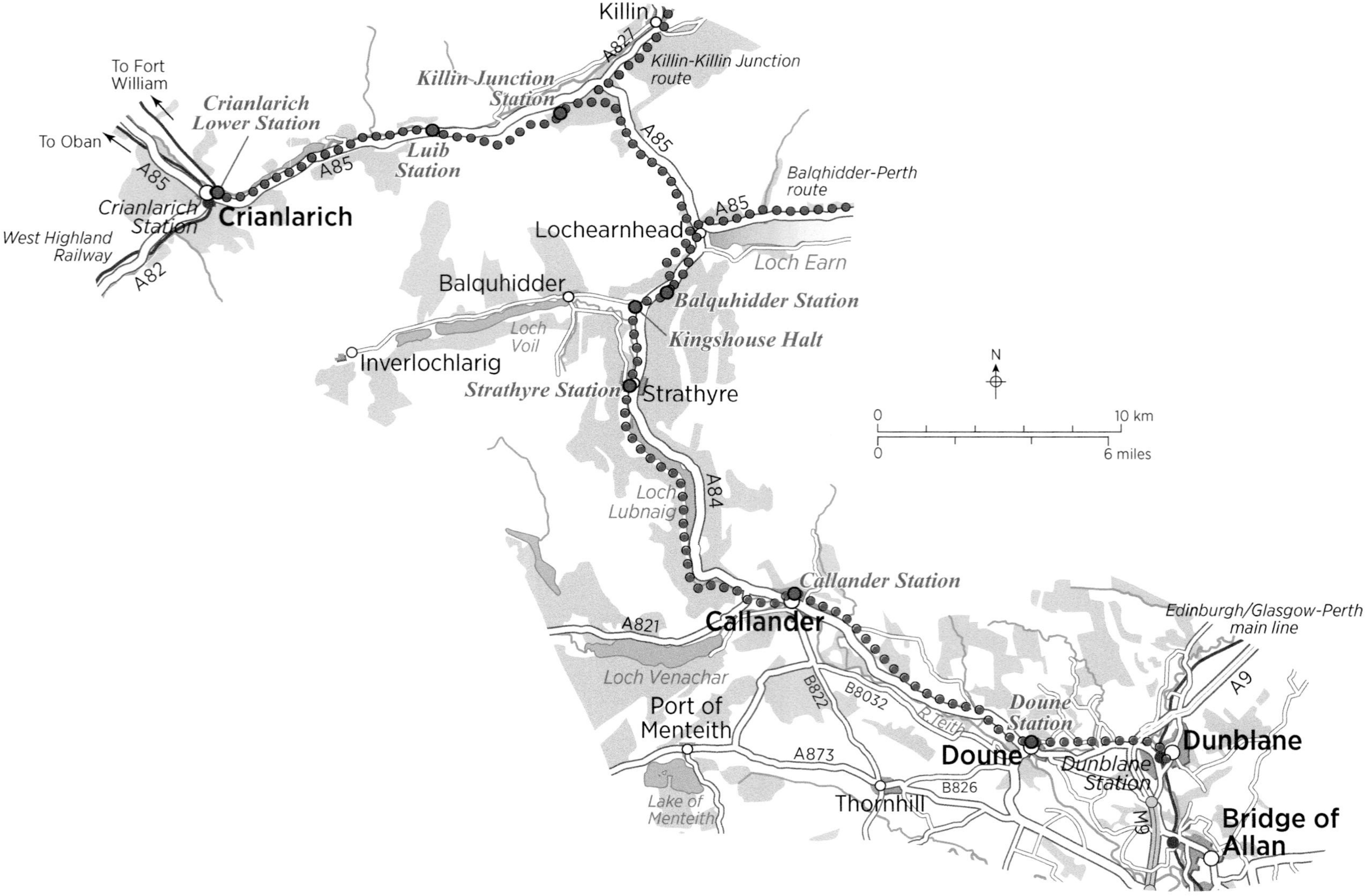

Killin
A827
Killin-Killin Junction route
To Fort William
Killin Junction Station
Crianlarich Lower Station
To Oban
Luib Station
A85
Balqhidder-Perth route
Crianlarich Station
Crianlarich
West Highland Railway
A82
Lochearnhead
Loch Earn
Balquhidder
Balquhidder Station
Loch Voil
Kingshouse Halt
Inverlochlarig
N
Strathyre Station
Strathyre
0
10 km
0
6 miles
A84
Loch Lubnaig
Callander Station
Edinburgh/Glasgow-Perth main line
A821
Callander
Loch Venachar
B822
B8032
A9
Port of Menteith
R.Teith
Doune Station
Dunblane
A873
Doune
Dunblane Station
B826
Lake of Menteith
Thornhill
M9
Bridge of Allan

Dunblane-Balquhidder-Killin Junction-Crianlarich

The track bed plus necessary detours offer a varied walk of over 40 miles, likely to be undertaken over several days. Map 57 covers the section from Dunblane to Balquhidder Station, map 51 the next stretch to Luib, and map 50 the final part to Crianlarich.

Dunblane to Doune station 4 miles

From the west side of Dunblane station, go past the Tesco store and immediately turn right to inspect a small section of track bed. Further on, steps on the right lead to the track bed/tarmac path; a short stretch only to Springbank Mill. Thereafter the track bed has been lost to housing developments as far as the A9.

Follow the A820 pavement over the A9, ignore cycleway signs and turn right, signposted Kilbryde. Pass by Stockbridge nursery to regain the track bed which may be muddy, waterlogged and cattle trodden. If so, continue west on the minor road to the bridge from where the track bed, grassy by now, is reached. This section, after a roughish start, offers excellent walking on a modest embankment.

The track bed goes under a bridge by Auchenteck farm but further on only one high pier remains of the viaduct over the Ardoch Burn. Follow the minor road to cross the burn and regain the track bed by a tarmac path, signposted National Cycle Network 765 and public path to Doune 1¾ miles. The tarmac track bed runs straight ahead, through an attractive tree-lined cutting, along an embankment (with on the right an old gravel quarry), and past an allotment, to reach the outskirts of Doune. Thereafter its line can still be discerned though all traces have been lost to urbanisation.

Now for a worthwhile detour. Doune (an Dun, the fort) is dominated by Doune Castle, situated on a promontory between the River Teith and the Ardoch Burn. One of the best-preserved 14th century castles in Scotland, its formidable battlements featured in the film Monty Python and the Holy Grail and more recently in the drama series Outlander.

Turn left into the village and follow the A820 to Station Wynd on the corner of which is Station House. No traces remain of the station at the top end of the Wynd. Nearby are the tranquil Doune Ponds and, as shown on the notice board, a path on the left leads to the A84 at the west end of Doune.

Doune to Callander station *8 miles*

Cross the A84 from where the track bed briefly follows the broad lane to Clarkton Farm. Although identifiable by the line of telegraph poles, it is then lost to the fields on the stretch to Buchany. Returning to the A84, the walk can be continued by the road-side pavement; however a more attractive way is to follow a 1¾ mile, partially signposted, public path by the River Teith, upstream to Old Kilmadoch Cemetery, then north by the Annet Burn to the western end of Buchany.

From the start of the Clarkton Farm lane, curve right then left on a track by the edge of a plantation. The winding path by the edge of woodland (and with open pasture on the right) curves round the high wooded banks of the Teith then descends to the river opposite Deanston. Head upstream, then slant over to a broad, grassy dyke-lined track, the old road to the long abandoned and ruined Old Kilmadoch church. Although only a small part of the east gable remains, the extensive burial ground is of considerable interest. Head uphill to a signpost, back the way, Doune 1½ miles, by the edge of pasture land. Follow the rhododendron-lined path to the broad wooded ravine of the Annet Burn. Pass by high piers, the remnants of the railway viaduct over the Burn, and so by path and gate to the A84.

Although all trace has gone of the railway bridge and abutments over the A84 at Buchany, the site is easily determined by the road`s curves and dip. (It is difficult to understand why the A84 has not been realigned.) Still extant is the original stone-arched road bridge across the Annet Burn.

A short distance west is the single-track road to Callander, signposted Drumloist. The road rises to the bridge over the track bed. To the right is a small cutting; to the left, the initial section between road and plantation is full of brambles and not worth the effort. Carefully follow the A84 side verge to regain the track bed by an entry point for vehicles.

Now tree-fringed on both sides, and easier walking, a lovely grassy stretch leads to the bridge access to Lower Coilentowie then into a small and possibly wet cutting. An overgrown minor embankment spans the Coillechat Burn and a weaving path leads to the Easter Coillechat access road. The railway bridge is no more but the far-side, tree-lined embankment is easily gained. Later the delightful grassy embankment, well seen from the road, bisects pasture land to reach the Wester Coillechat track.

Beyond that, the embankment continues through a wooded strip and then a wet-at-times grassy cutting to approach an incongruous building. This is in fact Drumvaich Crossing Signal Box. Although the Dunblane-Doune line was doubled in 1902, Doune-Callander remained single. The broad area in front of the box marks a crossing loop. The now restored box is a private residence.

Further on, the track bed is blocked by head-high overgrowth, requiring a short detour to the left, following a track round the eastern rim of a gravel pit. Regain the track bed at a padlocked gate which bears a warning about quicksand.

A long straight stretch between moorland and forest leads to the Keltie Water ("river" would be more apt), still spanned by the stone-arched railway viaduct. A mile upstream, and spanned by a footbridge, are the Bracklinn Falls (best visited from Callander) where over the ages the fast flowing Keltie, in a series of dramatic waterfalls, has cut a high-walled gorge into the hillside. Downstream from the railway viaduct are first the remnants of the old road bridge and then the concrete bridge carrying the A84; a mile further on, the Keltie joins the Teith.

The bridge over the road to Keltie Bridge Caravan Park is no more but the line of the track bed, and Cycle Route 76, is regained on the 2½ miles to Callander. A pleasing tarmac path descends gently to the outskirts of the town before urban development blocks the way. Follow the bicycle-signed route along Livingstone and Murdiston Avenues and so to a small car park, the site of the original Dunblane, Doune & Callander station at the then eastern end of the town. It became a goods yard soon after the Callander & Oban Railway opened to Glenoglehead.

At this point the cycleway goes up to the right. However, continue on a lane beside the old platform. At the end of the lane, head up a stepped path to Ancaster Road and then descend over the J Cameron Arrol girder bridge to reach what was the re-sited station, now a large public car and bus park. (Vehicular access is by Station Road, adjacent to the 1802 Dreadnought Hotel.) The former railway has been obliterated, on the west side by a grassy ramp and on the east side by houses.

J Cameron Arrol and his brother were successful builders of iron and steel structures at the turn of the 19th and 20th centuries. They became leading manufacturers, among other things, of Ferris Wheels. Their most striking structure is the giant wheel still in use in Vienna`s Prater Park. Rebuilt following World War Two damage, it featured in the classic film The Third Man....surely on everyone`s lists of films to see and places to go.

Callander to Strathyre station 9 miles

From Strathyre the River Balvag flows quietly into the 3-mile long Loch Lubnaig. However, the loch`s outflow, the Garbh Uisge, gathers speed as it enters the Pass of Leny and tumbles over the Falls of Leny to be joined by the Eas Gobhain, the outflow from Loch Venachar, to form the River Teith. That makes a lot of water flowing on the south side of Callander, so it is little wonder that floods occur there. Nevertheless, the height difference between Strathyre and Callander is less than 100m and, given the gentle gradient of the railway line, the walking is easy.

Heading west from Callander station and along part of National Cycle Network Route 7, pass by the slight evidence of a platform. Briefly follow Tulipan Crescent to cross the A84, and then a path descends to the track bed on a modest embankment towards the River Leny, nowadays spanned by a narrow pedestrian bridge. Little is left of the railway viaduct.

Follow the embankment which curves round the flood plain of the river. After 1¼ miles reach a parking area and a triple blue sign, Brig o`Turk 6 miles, straight on Strathyre 8 miles, and back the way to Callander. Cross with care the A821, the Loch Venachar road.

The railway kept to the west side of the narrowing gorge, the Pass of Leny, until steepening ground forced a crossing of the Garbh Uisge. A short diversion leads to the easily identified abutments but the bridge is no more. Continue on the path which hugs the steep wooded west bank in a series of tight turns and so to the southern end of a car park and the point where the railway re-crossed. Again only the abutments remain. (The Pass of Leny viaducts, as originally built, limited the classes of locomotives employed on the Oban line; weight restrictions were later eased.)

The car park, reached by a Bailey bridge across the river, is often busy; for to the west is the popular Corbett, Ben Ledi. Strathyre lies 6 miles distant. Regain the broad tarmac track bed, in effect a road through more open country by the quiet outflow from Loch Lubnaig. On the east side of the tranquil loch is the busy A84; a twisty road where safe overtaking opportunities are few and far between and just one slow moving vehicle can

Balquhidder Junction: ex-NBR Class D34 4-4-0 'Glen Douglas'(restored to NBR livery as Class K 256) heads a railway societies' special *D.Ferguson*

cause frustration and angst. The contrast with the west side of the loch is all too apparent; the delightful and peaceful track bed gives a relaxing walk or cycle. The loch is always beautiful, but especially when its waters are calm and reflective......a definite photo spot. Pass by Strathyre Cabins, beyond which the tarmac road comes to an end, replaced by a well-made path on the track bed in the shadow of the steep and craggy wooded slopes of Ben Ledi. Here rock-fall protection signals were installed—a more modern version of the elaborate arrangements in the Pass of Brander, which remain in being on the surviving western portion of the Callander & Oban route.

Reach a point where the path climbs away from the old railway. Despite the ascent, walkers and cyclists will stay with the path which later descends and curves right to rejoin the obvious grassy track bed. At a sign, Strathyre 1 mile, the path again detours in a steep zigzag ascent to join a forestry track which then descends to become a tarmac road and so to a sign, to the right Strathyre 400m, or, straight on Strathyre ¼ mile. Turn right, then left, for the village centre, and rejoin the track bed.

However, the intrepid track bed explorer will discard these easy detours. Close by the edge of Loch Lubnaig, the track bed in parts has seen much better days; yet there are sections offering a charming and easy way. This is what track bed exploration can be like!

Beyond a forlorn bridge, a stream has taken over and a large tree blocks the way, best avoided by dropping to the shore line. Eventually after a few wet and overgrown sections, like an explorer emerging from the jungle, stumble across the cycleway which is then

followed for a short distance. The second stretch of track bed is an easier affair. A cutting, once spanned by a bridge, can be detoured to the right. Later, with a few gates to cross, the straight tree-fringed formation gives an attractive, hidden way. Only latterly, around Kipp, may the way be muddy.

The track bed crosses the Balvag River by a wooden suspension bridge (the abutments of the railway bridge can still be seen) then continues by the Forestry Commission`s attractive sylvan car park to reach the site of the station at the southern end of the village. Celebrated in the popular song, Bonnie Strathyre, the village became a popular tourist spot in the 1870s following the arrival of the railway. In the 1890s it won the accolade of "best kept station" and was awarded a commemorative fountain with an elegant heron. When the station closed in 1965 the heron apparently "migrated" to a nearby garden.

The only evidence of the station is the now filled in overbridge on the northern side. Old Station Court, beside the village shop and car park, is of recent origin.

Strathyre to Kingshouse Platform 2 miles

Running between the River Balvag and the A84, the excellent new path between Strathyre and Kingshouse Platform mainly follows the line of the railway. However, amid the area transformed by the realignment of the A84, no trace of the halt remains nor of the point where the track bed crossed to the east side of the road.

Kingshouse to Balquhidder Station 1 mile

Using the Balquhidder village road underpass, carefully follow the grass verge on the east side of the A84 until opposite a parking spot. Head diagonally right and seek out the gap in the fence that leads to the track bed. A weaving grassy tree-fringed path (a short local walk by embankment and cutting) offers a secluded way to the station until housing and an overgrown cutting block the way. At the high metal bridge that carries the track from Leitters, follow the path on the left and then descend to the A84.

The station, opened as Lochearnhead in 1870, was renamed in 1904 and relocated further south to become the junction with the Comrie, St Fillans & Lochearnhead Railway. Behind the massive retaining wall the site (today occupied by Balquhidder Braes Holiday Park) was enlarged to include an island platform with subway entry, now boarded up. The outside steps (they are slippery, warns a notice) remain in situ.

Balquhidder Station is still so mapped. Balquhidder village lies two miles to the west.

Balquhidder Station to Glenoglehead Crossing 6 miles

Much of this section is utilised by National Cycle Network Route 7 and by the Rob Roy Way, the 79-mile long distance route from Drymen to Pitlochry.

From what is now the entrance to the holiday park, the Callander & Oban line curved left to cross the A84; but all evidence of the bridge has gone, the road has been realigned, and the far side has been occupied by houses. Carefully cross the road, head north for a short distance, then turn left down a lane to meet the cycle route on the far side of the embankment.

Turn right for a short distance to where a large section of the embankment has been removed to facilitate the Kendrum Hydro scheme. (At this point the cycle route can be followed, briefly on the Strathearn railway track bed, to a junction where a path leads uphill by zigzags to the Oban line through Glen Ogle.) Follow the broad track on the left to regain the track bed on an impressively high embankment. Later a cutting leads to open country and so to the Kendrum Burn bridge. Thereafter the track bed is used for estate purposes. The one ill-drained cutting is easily traversed, as is a short overgrown section. Rejoin the cycleway and the start of the glorious track bed tarmac path through Glen Ogle, arguably the most impressive section of railway construction in Scotland.

It is impossible not to be in awe of the cuttings and multi-arched viaducts required to obtain the requisite gradient on the ascent to Glenoglehead Station, nor be unaware of the recurring rock falls which required constant vigilance. We should be grateful to have inherited the dismantled line in the guise of a wonderful walking/cycling route.

The Callander & Oban Railway is terraced on the mountainside above, climbing into Glen Ogle *L.A.C. Horne, courtesy D. Ferguson* (see also p. 192)

The descent to Lochearnhead was, from the first, treated with respect, as the following extract from the 1881 Caledonian Railway working timetable discloses. ....*guards must put on their brakes, but not so hard as to skid the wheels or fire the blocks, and keep them on until the bottom of the incline is reached. Immediately after the guards have applied the brakes, the engine driver will apply what additional power is necessary and control the train with the brake power at his command. Engine drivers passing round 10 or 12 chain curves* [i.e. tight curves] *where there are check-rails must not exceed a speed of 12 miles an hour.*

This predates the universal requirement for continuous (automatic) train brakes.

With steep rock-strewn slopes to the west and open views to the east, it is a magical stroll uphill to the almost hidden Lochan Lairig Cheile and so to the well-preserved site of Glenoglehead station (mapped as Glenogle Cottages) which served as the original Killin Station until the opening of the branch line in 1886. Killin was subsequently reduced to a passing place.

Glenoglehead Crossing to Killin Junction and Luib Station 6 miles

Beyond Glenogle Cottages, the lovely broad tree-fringed track bed crosses the outflow of Lochan Lairig Cheile and gently descends, gradually curving west away from the A85. With many a bridge and embankment, the line is an impressive piece of work and, especially

Killin Junction: ex-LMS Black 5 4-6-0 45359 heads a 1950s west-bound goods train *Friends of the West Highland lines, courtesy D.Carmichael*

through the forest, gives easy walking. There are periodic views to Ben Lawers, and later over Glen Dochart. So to Killin Junction where platforms and siding, though overgrown, are readily identifiable; as at remodelled Balquhidder Junction, the layout was side-and-island.

(The branch line to Killin is described in another chapter.)

Pass by rhododendrons, a sign of an old estate, and then by an embankment to the remote viaduct over the Glendhu Burn. Though not so commanding as the viaducts in Glen Ogle, this structure is still impressive and is best appreciated by walking up a farm track on the left. Clear the forest by a wooden gate (dogs now on lead), pass under an accommodation bridge, then cross a wooden barrier. Now grassier and wetter at times, the track bed is still very walkable. Cross the Ledcharrie Burn viaduct where the line intersects the Kirkton Glen--Glen Dochart right of way.

A lovely grassy embankment, lined by silver birches, extends the walk to the Luib Burn. (A latter section is best avoided by climbing to the right by the fence posts.) Although the Luib viaduct is no more, crossing the burn dry shod is usually possible in summer and, unless in spate, a bare foot crossing at other times should do the trick. It is then an easy one-mile stroll by modest embankments and cuttings through open moorland to Luib Station, overlooked by the Munro Sgiath Chuil. Like Strathyre, Luib was a passing place with side platforms. Now the extensive site of Glen Dochart Holiday Park, the area has been landscaped and only the odd platform edge betrays the site`s history.

Luib to Crianlarich Lower station 6½ miles

From the west end of the Park, the grassy, silver birch-fringed track bed leads towards the noisy A85. A missing bridge over a lane necessitates a diversion to the road, which is then followed for a few yards before regaining the track bed. Reach a gate where the railway crossed the old A85. Road realignments have removed all traces of the bridge but on the north side of the A85 the track bed is again apparent.

The track bed then heads west for 2 miles between the road and the River Dochart. On a wet day, a few sections may be churned up by cattle. Cross the access track to Tigh an Fhraoich and reach the Allt Coire Chaorach which drains from the eastern slopes of Ben More. Luckily the unmapped sapling-covered bridge is still extant. A high embankment leads to a cutting spanned by a bridge. Turn right to avoid the cutting and follow the line of fence posts to regain the track bed.

Follow a winding path and cross the private road to Auchessan. By another short charming section, approach the spot where the railway crossed to the south side of the road. This cannot be determined exactly (road realignment) and, when the track bed path comes to an end, follow the grassy verge on the north side of the road to the picnic parking site by the shores of Loch Iubhair, *loch of the yew trees*.

(It is possible to cross with care to the south side of the road and follow the track bed for a short distance; ballast by the road side marks the approximate spot where the railway re-crossed to the north side, circa map ref 419262. Again no trace remains.)

Follow the road side verge to where the old A85 alignment emerges; back the way is a sign, parking space ¼ miles. Do not follow the old A85 but head right over a grassy area to the obvious track bed.

Crianlarich Lower timber siding and Glen Dochart (from WHR viaduct) *Crawford*

Now with both old road and new on the left, continue on the broad grassy embankment to reach a gate and a lengthy hedge where a detour to the left is required. Rejoin the old road and then turn right to cross the Allt Coire Ardrain by a small wooden bridge; very handy as only the abutments of the railway bridge remain. Pass by the white buildings of Portnellan and follow the signed path for Crianlarich.

The track bed goes by the shallow and partially reed-covered Loch Dochart. On a small wooded island is a ruined castle. Built by Sir Duncan Campbell between 1583 and 1631, only to be burned down in 1646, the building was more a three-story tower house than a castle. "A place of retirement and peace" opined Dorothy Wordsworth, and certainly the loch-side way gives a most pleasant and relaxing stroll. Later a slight embankment between road and loch offers a beautiful, winding tree-fringed walk beneath the bulky slopes of Ben More.

Ahead lies the end of the walk. Cross a stile to continue on a broad grassy way, also used by cattle, to reach the whitewashed Inverardran Cottage (Ochils Mountaineering Club). The track bed is then lost to overgrowth and modern buildings, so turn left to the roadside pavement by Ben More Lodge Hotel.

Though an obvious transport junction (and an early target of railway promoters) Crianlarich remains little more than a village; yet it came to possess two stations, Lower and Upper. Crianlarich Lower, ex Callander & Oban, became for some years after closure a timber terminal for trains feeding the Corpach pulp-mill at Fort William. The site is now occupied by a car park, a community centre and a fire station. No station buildings remain. To the west can be seen the West Highland Railway viaduct. From the Upper station, no longer differentiated, today`s trains head north to Fort William and Mallaig, or turn west for Oban.

Killin Junction-Killin

The Killin branch (*strictly* Killin Railway) extended to Loch Tay, giving steamer connection to Kenmore. (Refer to Aberfeldy-Ballinluig Walk.) Sailings, which in their day had provided a passenger and cargo service, ceased in 1939. However, the little line`s pier extension remained in use—to access the engine shed. The Lochay hydro-electric scheme of the 1950s, with the power station sited at Finlarig, gave a temporary boost to the railway`s inevitably declining traffic.

Like Riccarton Junction on the Waverley Route, Killin Junction, map ref 533291, was accessible only by rail. Purely an exchange station with relatively few trains, it never developed into a railway community and operating centre. For today`s walker that means starting from Lix Toll or from the head of Glen Ogle. (Refer to Dunblane--Crianlarich Walk.)

Map 51 covers the 5-mile walk.

The drop in height from Killin Junction to Killin Pier is 130m; a 5-mile gradual descent that gives easy walking. The initial track bed is now in effect a forestry track; a secluded way through extensive and dense woodland with the nearby A85 mostly out of view. The track

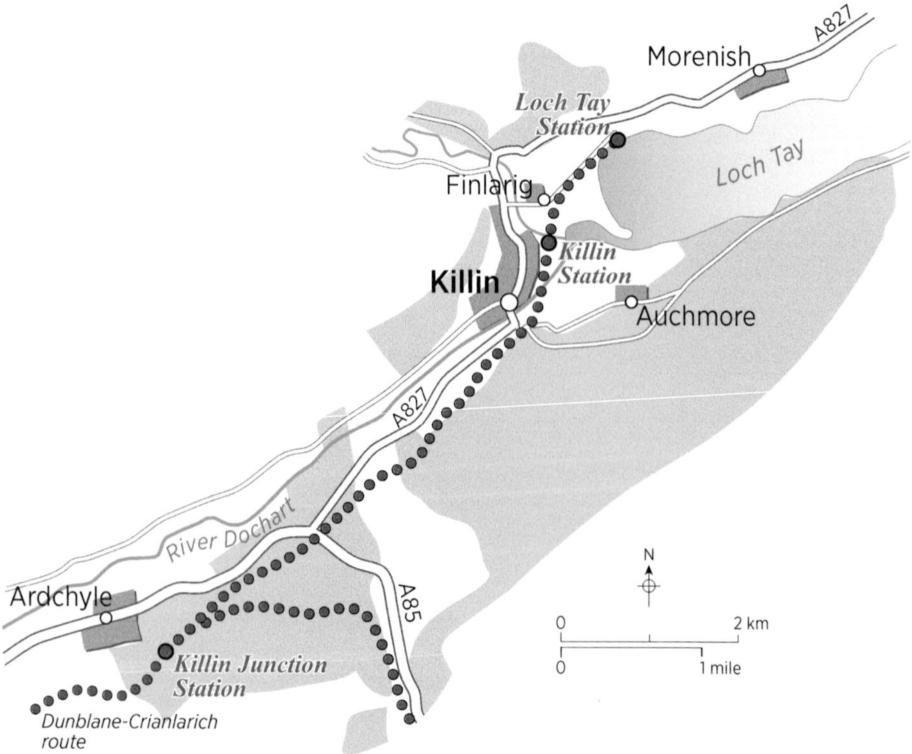

Killin Junction, converging track beds *Crawford*

slants down to a line of pylons and on to Lix Toll. The Roman numerals for 59 (LIX), recalling the 59th Legion, is one fanciful explanation for the name which is more likely to have derived from *leac*, Gaelic for a slab or gravestone, or from *laxa* of Scandinavian origin meaning salmon.

Some 100 yards south of Lix Toll road junction, the railway bridge over the A85 is no more but the spot can be identified by the west side embankment. Cross the road with care as bends do not give long sight lines. A few yards uphill regain the track bed on a slight embankment, where a winding path gives a charming stroll. (On my day two fallen trees posed no great problem.)

Cross a burn and so to where Cycle Way 7 comes in from the right then utilises the track bed. This becomes a broad, grassy-sided way between silver birches, ultimately leading to the outskirts of Killin. (At one point a stack of redundant concrete sleepers testifies to the railway past.)

Gardens and houses bar the way, as does a missing bridge over the road that leads to the south side of Loch Tay. Turn left as signed to the A827 and pass by the scenic and much photographed Falls of Dochart.....cue for a photo stop. Do not cross the Dochart by the road bridge. Follow a track, tarmac at first, close by the gorge and parallel to the old railway, to reach an overbridge spanning the track to Kinnell. Then keep to the left to regain the track bed.

A high embankment leads to the A-listed Dochart Viaduct, one of the earliest mass concrete railway structures in Britain. The viaduct is 37ft high and the five 30ft-span arches on masonry piers were each built in a day.

The track bed skirts the east side of Killin and a minor detour is required. Pass the sewage works and so to a car park, bus stances and highway maintenance depot; the site of Killin station, of which no trace remains.

The final stretch to Loch Tay Station, a very popular local walk, can claim to be the most beautiful one mile stretch of any track bed. On a modest tree-fringed embankment over what is often little more than the flood plain of the swollen River Lochay, it is a charming secluded way overlooked by the craggy Meall nan Tarmachan. I first visited this area during a very wet spell, when crossing the iron bridge over the river led to an "unmapped" western expansion of Loch Tay, such that the track bed was the only ground above water.

The track bed path then goes by the shingly shores of Loch Tay and so to journey`s end at three gates...on the left Finlarig Lodge, to the right Pier End and in the middle Cory End. Cory is an acronym for Callander & Oban Railway Company. Kry End (the terminus of Killin Railway) lacks appeal!

On the right of the track bed line to Cory End is the old wooden station building. Further on can be seen Finlarig power station. The hydro scheme has the highest head (at 415m) of any in the UK. As for the old steamer pier, only traces of the foundation piles can be seen at times of low water. Today`s small wooden pier is of later origin.

Balquhidder -Crieff- Perth

The Caledonian Railway and their LMS successor offered late afternoon/evening "circulars" in summer from Glasgow and Edinburgh via Gleneagles, Crieff, Loch Earn and Callander. A similar service pre1914 was the Strathearn Express giving summer residential businessmen a convenient through service to the cities, returning in the evening. As with the North British Company`s Fife Coast Express when first conceived (refer to Leven—St Andrews Walk), the idea was to tap those middle class families who took summer quarters.

The Strathearn track beds, varying from superb to impassable (with some sections urbanised or ploughed up), offer the walker a challenging and exhilarating 31-mile traverse, albeit with gaps, from Balquhidder Station to Perth. From this writer`s perspective, exploring the line was frustrating at times yet hugely rewarding in uncovering hidden gems. If, regretfully, the reader has only time to explore one line, then arguably this is it.

Two Ordnance Survey maps are required, 51 and 58 (and briefly 57).

Balquhidder Station to Lochearnhead station 3 miles

(Also refer to Dunblane-Crianlarich Walk.)

From the northern end of the station site, the track bed continues for one mile parallel to the A84. A rougher overgrown area is easily followed via a weaving path. Reach a gate and a lovely cleared section of track bed which gently descends, curving towards the road.

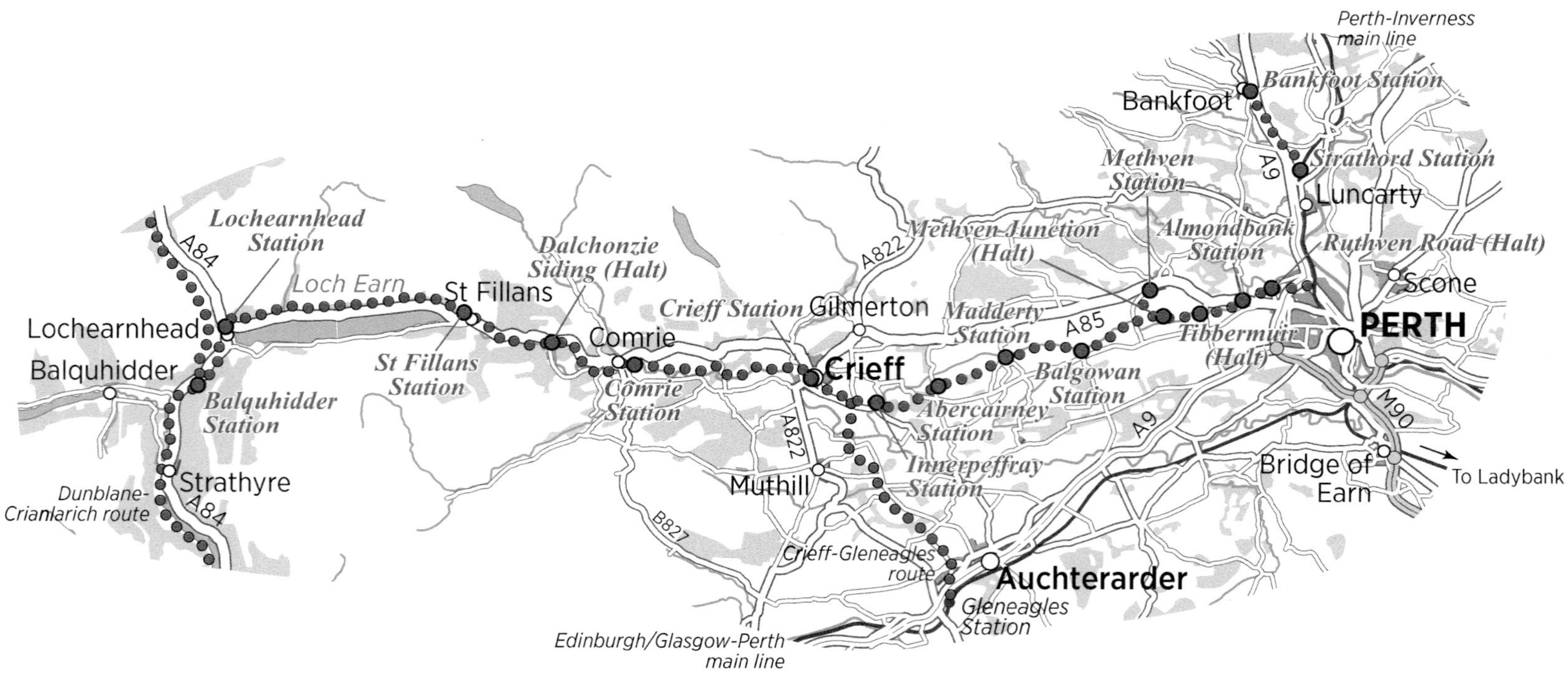

Perth-Inverness main line
Bankfoot Station
Bankfoot
Strathord Station
Luncarty
Methven Station
Almondbank Station
Ruthven Road (Halt)
Scone
PERTH
Methven Junction (Halt)
Madderty Station
A85
Tibbermuir (Halt)
Balgowan Station
Abercairney Station
Innerpeffray Station
A9
M90
Bridge of Earn
To Ladybank
Lochearnhead Station
A84
Loch Earn
Dalchonzie Siding (Halt)
St Fillans
St Fillans Station
Lochearnhead
Balquhidder
Balquhidder Station
Strathyre
Dunblane-Crianlarich route
Comrie
Comrie Station
Crieff Station
Gilmerton
A822
Crieff
Muthill
B827
Crieff-Gleneagles route
Auchterarder
Gleneagles Station
Edinburgh/Glasgow-Perth main line

Kendrum viaduct *Crawford*

Continue past the high padlocked gate on the left, the exit point for the estate, and stay on the modest embankment to the point where a bridge spanned the A84. No evidence of it remains.

Scramble down to the road and cross with care to the far side. Beyond a small fence is Cycleway Route 7 which within a few yards rises to the track bed embankment. Ahead lies a spectacular section, the ravine of the Kendrum Burn spanned by the 1905 7-arch Kendrum (or Edinchip) Viaduct, 100ft high and difficult to photograph.

Today, pedestrians and cyclists can cross courtesy of the new central span; financed by funds raised in memory of Nigel Hester, organist and music teacher, who was killed whilst cycling on the A9. Without the old solid parapets, it is an airy crossing.

A gentle descent on the receding embankment leads to an overbridge and a signposted junction. (Once under the bridge, the cycle way path zigzags uphill to the left to join the Glen Ogle track bed.) The immediate track bed towards Lochearnhead goes through an overgrown and wet cutting....cue for a short detour.

As signposted Lochearnhead ½ mile, head right before the bridge and follow the concrete path to reach what is the top end of Craggan Road. Continue downhill for some 100 yards then cut up to the left to the rim of a cutting. Follow a worn path to regain the track bed. A weaving path gradually develops on the slight embankment by the back of houses. Pass under a bridge and ahead is the first sight of the broad island platform of the charming Lochearnhead station.

Opened in 1979 by Hertfordshire Scouts and now known as Lochearnhead Scout Station, the converted buildings (still bearing advertising boards), platforms and surroundings, have been immaculately preserved and maintained as an activity centre. The subway to the village is now blocked by houses.

The station access road leads to the A85, where only the eastern abutment of the railway bridge remains. Climb to the far side embankment and the start of the 9-arch concrete viaduct that spans lower Glen Ogle. Although the far end is blocked by a fence then housing, the short out-and-back walk is recommended.

Lochearnhead to St Fillans station 7 miles

On this section, construction of the railway involved cuttings, embankments and one short tunnel, as well as viaducts over the numerous side streams. Some of the cuttings are overgrown and may be wet. All these difficulties, however, can be detoured. The track bed, parallel to an electricity transmission line, offers a surprisingly varied walk, partly through a wooded area intermittently masking both Loch Earn and the A85.

As of summer 2017, part of the track bed from just west of St Fillans (and east to where it passes under the A85 at Tynreoch) has been upgraded to a tarmac cycleway/path; a community based effort known as the Loch Earn Railway Path (LERP). The upgrade includes replacing the missing central span of the Glen Tarken Viaduct and opening the previously gated short tunnel.

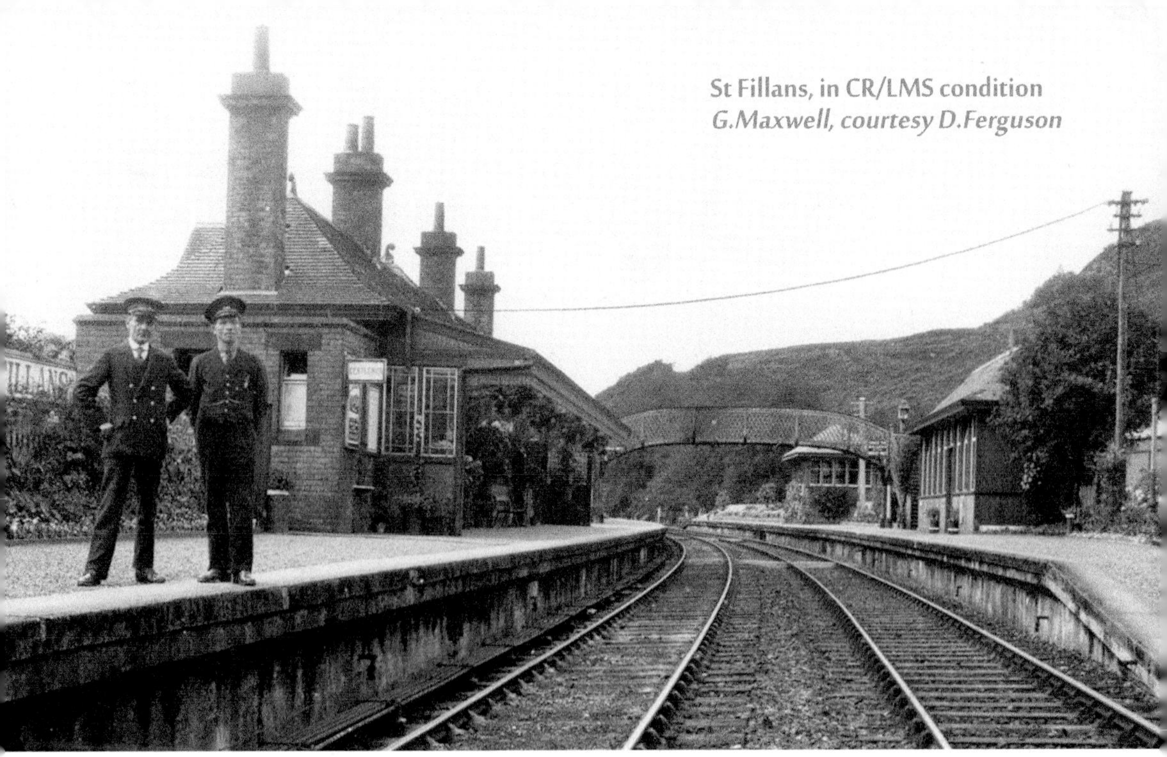

St Fillans, in CR/LMS condition
G.Maxwell, courtesy D.Ferguson

Presumably it is the intention to eventually complete the LERP all the way to Lochearnhead, perhaps with a tarmac cycleway/path for some of the distance. For those cyclists who inexplicably do not like the quiet road on the south side of Loch Earn, nor (understandably) the busy and dangerous A85, the attraction of a designated almost hidden route is all too obvious.

For walkers, subject to a few diversions, the whole distance is easily traversed and those explorers who may be less enthused about any "improvements" should do the traverse from west to east before the LERP is completed.

From the car park in Lochearnhead, go up Auchraw Terrace; turn right on a track to reach a house on the left. At this point a grassy way zigzags to a railway bridge over a very deep, wet and overgrown cutting. Return to the track which climbs to the formation beyond the cutting.

At first a long straight line above the village, the track bed becomes a grassy way. Where a farm shed and a wet cutting impede progress, take the path on the south side. Later there are lovely views over Loch Earn to Ben Vorlich and Stuc a`Chroin. The next obstacle, an overgrown cutting, requires a detour to the left to regain the track bed.

A missing bridge span requires another detour, after which a grassy embankment is followed to the Beich Burn viaduct, overgrown with saplings but still walkable. The next section, an estate track through Ardveich and Derry Woods, is more terraced into the hillside. It would be a shame if this delightful stretch were "improved" by a tarmac cycleway/path. Beyond a large turning area, the track bed becomes a lovely winding way through a slightly overgrown area.

At Glen Tarken, from where a side track gives easy access to the A85, is a triple sign, Lochearnhead 5 miles; straight on St Fillans 1¾ miles; and to the left, a longer route via Glen Tarken, St Fillans 6 miles.

A lovely new path through a former very wet cutting (what an improvement!) leads to what must have been an impressive viaduct high above the Tarken Burn. The missing central span has been replaced by a new wooden-decked girder bridge, thus obviating any ill-advised scramble into the deep gorge.

There are many side streams and gorges on the approach to St Fillans through Glentarken Wood, but luckily all are still spanned by bridges. High above the road and with St Fillans only ¾ miles distant, a lovely, grassy, silver birch-lined narrow path eventually reaches the new section—a broad way which has transformed the approach to the line`s lone tunnel, previously barred by a high metal padlocked gate. The bore is short, hence no torches are required. It is goodbye to the previous detour path.

Bypassing a house built on the track bed, the broad way curves right to Station Road. Ahead is the entrance to the site of St Fillans station; a large area now occupied by St Fillans Caravan Park, where chalets and static caravans surround the well-preserved platforms. Do courteously request a visit.

St Fillans to Comrie station 6 miles

There is a bus service from St Fillans to Comrie, operated by Stagecoach Perth—Traveline 0871-200-2233

Return to Station Road. Head uphill by the abutments of the rail bridge and turn right on the new path, signed Dalchonzie 3 miles and Comrie 6 miles, that detours the caravan park and a waterlogged cutting. A wooden bridge leads over a burn to the tarmac cycleway/path which zigzags uphill then descends to the track bed. Note to the south the pointed peak of Am Bioran and pass on the left the brightly coloured Crocodile or Serpent Rock, first painted by youths over 100 years ago. The rock is said to lie on a ley, a supposed prehistoric straight "energy" line usually between hill tops.

A small cutting leads to the A85 underpass by Tynreoch. The track bed, now a forestry access road, crosses the River Earn by a girder bridge and heads through open country towards Dundurn Forest beneath the craggy slopes of Mor Bheinn. Just beyond the bridge, where it becomes a grassy farm track, is a gate with a sign.... "Take care, livestock including cows and calves are often found in this area and a bull may sometimes be present. Do not take dogs into this field when young farm animals are present and keep dogs on a short lead

or under control at all times." So much better than the old "beware of the bull" sign! There were no animals on my day. At the far end of the field is another sign, Dalchonzie 1 mile.

Traverse a lovely wooded section, cross the Earn again by a bow-string bridge then follow an impressive rocky cutting and so to a padlocked gate. Use a side gate and cross a minor road to the site of Dalchonzie Platform where the signal box and adjacent building have been converted into a house.

Further on, the track bed round the edge of Drumlochlan Wood is completely overgrown and the parallel drive only leads to private housing. Moreover, the next bridge over the Earn has been dismantled. But no matter: the minor road on the south bank, rather than the A85, gives a pleasant two-mile stroll to the charming stone-arched Ross Bridge (1792) at the west end of Comrie. Turn left on Station Road, whence path and track bed lead to the pedestrian bridge over the River Lednock, at the east end of the village. A tarmac road by the line of the vanished railway leads to the A85. The overbridge has been filled in and on the far side is the site of Comrie station which has suffered the fate of many in becoming a holiday park.

(The ex-Caledonian line should not be confused with the Comrie Railroad which opens to the public but only on special occasions.)

Comrie to Crieff station 6 miles

To avoid the caravan park, turn east on the A85, pass Tomperran and continue to a farm lane, map ref 782225, beyond which the former railway is regained. Now on a modest

embankment, the charming tree-lined track bed parallel to the Earn, and well away from the road, offers good walking. Later, close by the river, the weaving path along the sapling covered embankment reaches the only impediment...a missing bridge where a burn has cut through the formation. However, a fallen tree and small planks make an interesting crossing. (The burn, where it joins the river, can be splashed over.)

Traverse a path winding through another overgrown area to clear the wood. Ahead is a lovely way (mapped as a track and now briefly on map 57) on a higher embankment, heading towards the prominent obelisk, built in memory of General Sir David Baird, 1757-1829. The trackbed curves into a rocky cutting and a short tunnel under the Trowan road. Despite the sound of dripping water, the tunnel appears to be in good order but the entrance gate is padlocked. Retrace steps and follow the road to the far end of the tunnel. Turn left by Trowan farm to regain the track bed.

On the approach to Crieff, the track bed is of variable quality. At one stage resort to the field edge and then enter a cutting with fallen trees. On the right, beyond a missing bridge, is Lady Mary`s Walk, a favourite of Lady Mary Murray whose family owned the land in the early 1800s. The broad attractive path, really more of a track, goes through a beautiful avenue of mature trees, mostly beech and oak, and parallel to the track bed. The inscription on one bench aptly sums up the contrast...."the tumbling bridge and fallow field, once deeply scarred by soot and steel, now nature has crept back to heal".

At a bifurcation (signed ¼ mile to Crieff), take the left path to rejoin the track bed at a cutting, after which an embankment leads to the Turret Burn. Only the stone pillars of the viaduct remain.

Go upstream, cross the road bridge over the Turret and immediately head south to overlook Morrison`s Academy playing field. It is then an interesting exercise to follow the line of the railway, cutting SE through the town. The first part can be identified by following a stepped rising path past a cutting and an old stone bridge. The line then disappears before reaching the Co-operative store and car park, the site of the station. At the east end is the Station Bar and a short stretch of cutting.

Further SE the line is lost to urban development, a school and leisure centre. The only remnant is a short stretch of track bed, now a lane, which gives access from Robbie`s Box (Pittenzie Crossing) to a builder`s yard, the site of the railway workshops. Beyond, the track bed has been ploughed up as has that of the parallel line to Gleneagles.

Innerpeffray station to Abercairny station 2½ miles

Innerpeffray station house, mapped as Peveryl and with a station plaque attached to the wall, stands in private grounds just north of Millhills. The site is clearly identified by a grassy embankment leading to the high abutments of the former bridge over the minor road. I was privileged in being shown around the beautiful lawn covered platforms.... but remember that the site is private. The track bed to the west, although still mapped, is overgrown and not walkable.

Head east and pass by the field edge on the northern side of farm buildings to reach an embankment and a large single-span stone-arched bridge that takes the track bed across the Pow Burn, which flows to the Earn. What follows is a lovely curving way giving open views. Also used by farm vehicles, this one mile section to the farm lane north of Dolleriemuir will pass all too soon.

Across the lane is the western end of Long Plantation. The line of the mapped track bed, now ploughed up, crosses a field, marked by a single telegraph pole; and unless there is a tractor way amid growing crops it will be necessary to make an admittedly pleasant diversion by lane and road past Tuchethill. If it is possible to cross the field, continue to rejoin the track bed. The northern side of the plantation has been felled. For vehicle access to the trees on the south side, the modest railway embankment has been cut in three places,

but these present no problem. Reach the remains of the bridge over the minor Tuchethill road, easily identified by the remaining abutments.

Although it is possible to continue, at first by the plantation edge, the track bed, particularly in the latter stage, becomes overgrown. It is better to make a triangular detour by road to map ref 922208 where the bridge spanning the track bed overlooks what is a very welcome surprise...Abercairney, the most readily recognisable of all the stations east of St Fillans. The lengthy stone platforms and the stationmaster`s house appear to be in a time warp, with the clock on the wall of the house stopped at 10.25.

Abercairny to Madderty station 3 miles

To avoid the private grounds at Abercairny, head south then east to Cowdens, from where a track regains the track bed; a farm access way which makes for brisk walking as far as the mid-point of the wood on the north side. Thereafter it is slower going, rough and grassy until the track to Ardbennie is reached. (Further on the formation is lost to the fields.)

From Ardbennie, follow the tarmac road that undulates towards Madderty. (A lane on the left leads to the white-painted Madderty church.) Pass Madderty school and descend, as if heading for the ruins of Inchaffray Abbey; on the right is Madderty station house where the lovely landscaped garden covers the site of the station.

The overgrown trackbed continues for two miles by the Pow Water, but is then lost before reaching Balgowan station. Given this rough and distinctly unappealing stretch, an out-and-back walk is not recommended.

Balgowan station to Methven Junction 2 miles

In Balgowan, follow the lane to the recycling point, the site of the station of which no trace remains. Follow a worn path by the Pow Water to regain the rough-at-first grassy track bed, adjacent to new houses. The going then improves. Cross a side track to enter the open expanse of the appealing Methven Moss. The track bed, at first well-used by vehicles, leads to a wooded area where one section may be slightly wet. However, most appropriately, it soon

Balgowan; station staff in CR uniform L.A.G.Thorne, courtesy D.Ferguson.

becomes a mossy way. Hidden, and fringed by silver birches, this short stretch is arguably the most attractive part of the Balquhidder to Perth Walk.

All too soon reach Mossgates Cottage, which housed the railway gatekeeper responsible for the level crossing on the road from Methven. Beyond the crossing the track bed has been ploughed up, as has all trace of Methven Junction, south of Mains of Tippermallo. Nothing remains of Methven station, now a small industrial site reached via Station Road. However, after crossing a burn, a pleasant short stretch of tree-fringed track bed, much used by local walkers, can be followed into Burnside.

Tibbermuir station to Almondbank station 1½ miles

All that remains of Tibbermuir station at Duntanlich, ½ mile north of Tibbermore, is a small remnant of a wall. (It is possible to head west, at first on the lane by houses, then on the rough but walkable grassy track bed. However, no trace remains of the bridge over the Pow Burn and further west the trackbed has been ploughed up.)

Heading east, a few houses block immediate access to the track bed. Seek permission to pass by on the south side, or squeeze by the narrow wooded strip between houses and burn, to reach a broad cutting that gives easy walking. Beyond the track from Marlefield, a tree-studded section with a weaving path leads to a modest embankment, lined by mature silver birches. Parallel to the Pow Burn and with fields on either side, this short section is second only to Methven Moss as the most secret and lovely part of the railway line.

On reaching an overgrown cutting, cross the fence to enter a partially wooded area on the right; the site of old mining and a popular local walking area. Head south-east to a broad lane and so to the minor road by North Blackruthven. This detour is required to avoid houses built on the site of Almondbank station at Lochty, of which all traces have gone apart from Station House. The A85 overbridge has been filled in.

The track bed east of Lochty has been lost to open fields and a small industrial estate. Just one short remnant survives by Ruthven House; a grassy then rougher section that leads to the M90.....an inglorious end to a glorious walk. The former Almond Valley Junction, on the Perth—Inverness main line, is out of reach.

Crieff - Gleneagles

Crieff Junction station, subsequently Gleneagles, became a much grander affair in conjunction with the 1924 opening of Gleneagles Hotel. Built for the benefit of the well-to-do, the hotel was to be reached exclusively by rail and indeed had its own little branch and private siding.

Between Crieff and Gleneagles were three stations....Highlandman, Muthill (mapped as Caerlaverock, two miles east of the town) and Tullibardine.

The line to Crieff had the bonus of providing tourist access to the Strathearn Hydropathic establishment, opened in 1868, nowadays Crieff Hydro.

Map 58 covers the 9-mile route.

Railbus at Crieff *L.A.C.Horne, courtesy D.Ferguson*

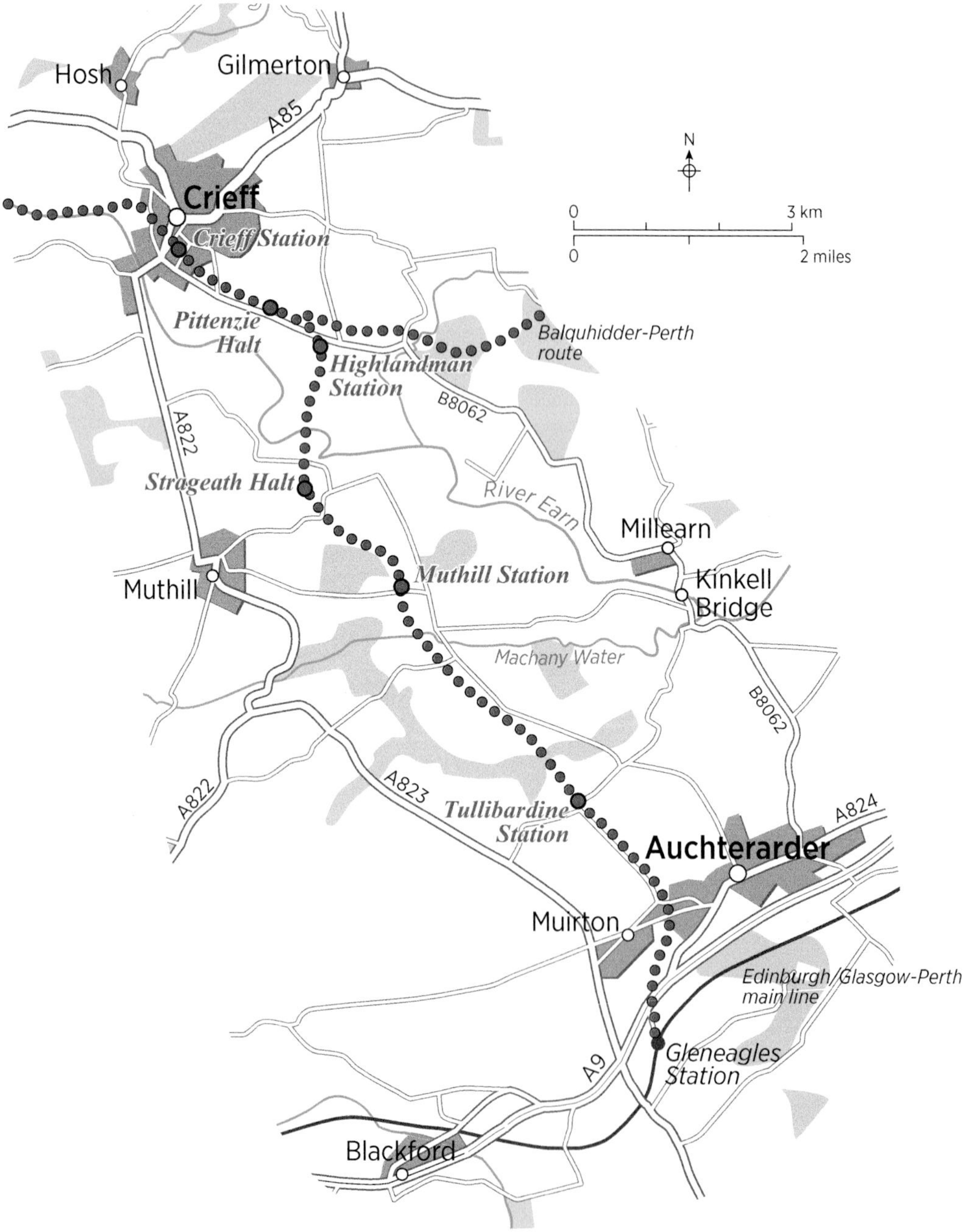
Hosh
Gilmerton
A85
Crieff
Crieff Station
N
0
3 km
0
2 miles
Pittenzie Halt
Highlandman Station
Balquhidder-Perth route
B8062
A822
Strageath Halt
River Earn
Millearn
Muthill
Muthill Station
Kinkell Bridge
Machany Water
B8062
A823
A822
Tullibardine Station
A824
Auchterarder
Muirton
Edinburgh/Glasgow-Perth main line
Gleneagles Station
A9
Blackford

Highlandman station

First a nostalgic visit to Highlandman station. At Crieff the railway ran parallel for one mile to the line from Methven (both at this point now subsumed by fields) then curved south to cross the B8062. Although there is no trace of the level-crossing, the station building on the south side of the road is easily identified by the large sign. A wall-mounted clock, stuck at 3.53pm, overlooks what is left of the platform. The overgrown track bed continues by the side of the small modern housing estate, at the far end of which was once a large sawmill that provided much timber traffic for the railway. The end house owner is pleased with his well-draining sawdust/clay soil! The track bed is then lost to the fields, and further south, though the piers remain, the bridge over the River Earn is gone.

Crieff to Muthill station 4 miles

At the southern boundary of the town, by the entrance to the Stuart Crystal Visitor Centre car park off the A822, is a green sign, footpath to Strageath via Templemill by the banks of the Earn. By Templemill farm is a footpath sign to the right, but carry straight on as

indicated, Strageath 1 mile; a particularly attractive stretch, to reach the massive piers of what was once the 4-span bowstring girder bridge.

Leave the riverbank and head south on the left hand side of the embankment, then climb to reach the track bed and so under a bridge. Continue in a straight line on the grassy way as far as the parapets of an old bridge beyond which the track bed is lost. Turn left by the track for Strageath Halt, only built in 1958 when an economical diesel railbus was introduced....an unsuccessful attempt to save the branch. Then turn right on the farm road to reach a minor tarmac road. Turn right, then shortly left by the signpost, to enter Sallyardoch Wood; at this point the public walk wends its way to the right through the charming woodland and so to Muthill.

Muthill *D.Ferguson*

However, slant left, quickly reaching the obvious track bed. With a new plantation on the right, the gently rising, easily walked, track bed gives open views to the north. A short rougher section is easily negotiated and ahead can be seen the large seed potato farming complex at Caerlaverock, the site of Muthill station of which nothing now remains.

Regrettably the final 50 yards of track bed has been enclosed by two barbed wire fences, meaning a retreat to Sallyardoch Wood. Nevertheless, despite having to retrace steps, the out-and-back section is still well worth doing.

It is a pity that negotiations between Perth Council and the landowner to establish a short bypass path have not been successful; the more so because such a path has been installed at Drumness, ¾ miles to the south.

Muthill to Tullibardine station 2½ miles

Cross the Muthill road to a new path, Drumness ¾ mile, that avoids a house and garden on the track bed. Regain the track bed, a lovely grassy way, which gently descends to the bridge over the Machany Water then rises to the farm complex at Drumness where the new path, courtesy of the farmer, keeps west of the farm buildings.

The railway crossed the immediate higher ground (note the nearby trig point at 86m) by passing under the minor road to a wide, deep rocky cutting. No trace remains of the overbridge and, although the impressive cutting is extant, it runs through the curtilage of Machany House. Time for a short detour.

From Drumness, turn right on the Tullibardine/Auchterarder road and then right on the track towards Farmton. Just before a filled-in yet easily recognised overbridge (with time to admire the Machany private grounds), turn left, and then right by a footbridge across a

burn, to regain the track bed. Pass through two gates to a grassy embankment and so to a bridge over a farm track.

By now a secluded way, not apparent from the map, the track bed gives a grassy and enjoyable stroll passing by a number of gates, some of the horse-rider friendly type. Only at one cutting, which may be partially flooded, is a careful traverse required. Reach a large wooden gate and a track coming in from the right.

By now mapped as a track, as indeed it is, the old formation clears the wood then, by a tree-fringed embankment, crosses a farm track and another burn. A steady rise leads to a widening grassy lawn and a large white modern house on the track bed....the site of Tullibardine station.

Although the way ahead is barred, a narrow path on the right and another on the left (once the station access) lead to the road. Note the sign on the ground, Old Station House. An abutment (on the station side) is all that remains of the bridge over the Tullibardine village road.

To the west is the atmospheric Tullibardine Chapel, founded in 1446 and one of the few medieval churches in Scotland that survived the Reformation. The walls are studded with the coats of arms of the Murray family.

Tullibardine to Gleneagles (Crieff Junction) 2½ miles

Although Gleneagles station lies 2½ miles distant, only the immediate ½ mile, signed Old Railway Path, can be identified as the track bed.

A ramp leads to the grassy embankment, parallel to the road to Auchterarder; a pleasing section as far as a National Grid storage installation. After this, as mapped, the track bed is subsumed by fields. Thereafter it is an urban though most pleasant stroll by the west end of Auchterarder to the entrance to Gleneagles Hotel.

At the western end of Auchterarder, no trace remains of the two overbridges. Further on, the spread of Gleneagles (golf courses and houses) and the construction of the A9 dual carriageway have completely changed the landscape and all but erased signs of the railway. Gone too is the spur to Gleneagles Hotel.

From the hotel entrance, head down old Station Road now bisected by the A9. A pedestrian way over the dual carriageway demands a most cautious crossing and so to the northern end of the station. From there, a very short stretch of the track bed, now a lane, can be identified; the point where the Crieff line gradually curved away from the Stirling—Perth line. After 100 yards leave the lane and follow a winding path through the wooded area until all signs of the track bed have gone. However, the path gives a short circular walk back to the station.

Aberfeldy-Ballinluig

Map 52 covers the 9-mile single track railway The only branch at the southern end of the Highland Company`s system, it stopped short of Kenmore and convenient connection with the Loch Tay steamer.

Aberfeldy to Grandtully station 5 miles

Aberfeldy station, on the east side of the town, has been subsumed by urban development. Although the first two miles of the track bed, on the south side of the A827, are still partly mapped (and identifiable from the road side), the short stretches that can be walked scarcely make the effort worthwhile. Instead follow the 5-mile waymarked Aberfeldy Riverside Walk (part of the Rob Roy Way) which stays adjacent to the road, then zigzags down to the Tay. Look out for canoeists and rafters paddling easily downstream on the meandering river.... that is until reaching Grandtully (pronounced Grantly).

The path goes through a lovely stand of silver birches to join the trackbed, by now on the north side of the road. (The bridge over the A827 is no more, though the site can still be identified.) Go under the bridge that carries the A827 to arrive at the readily identifiable

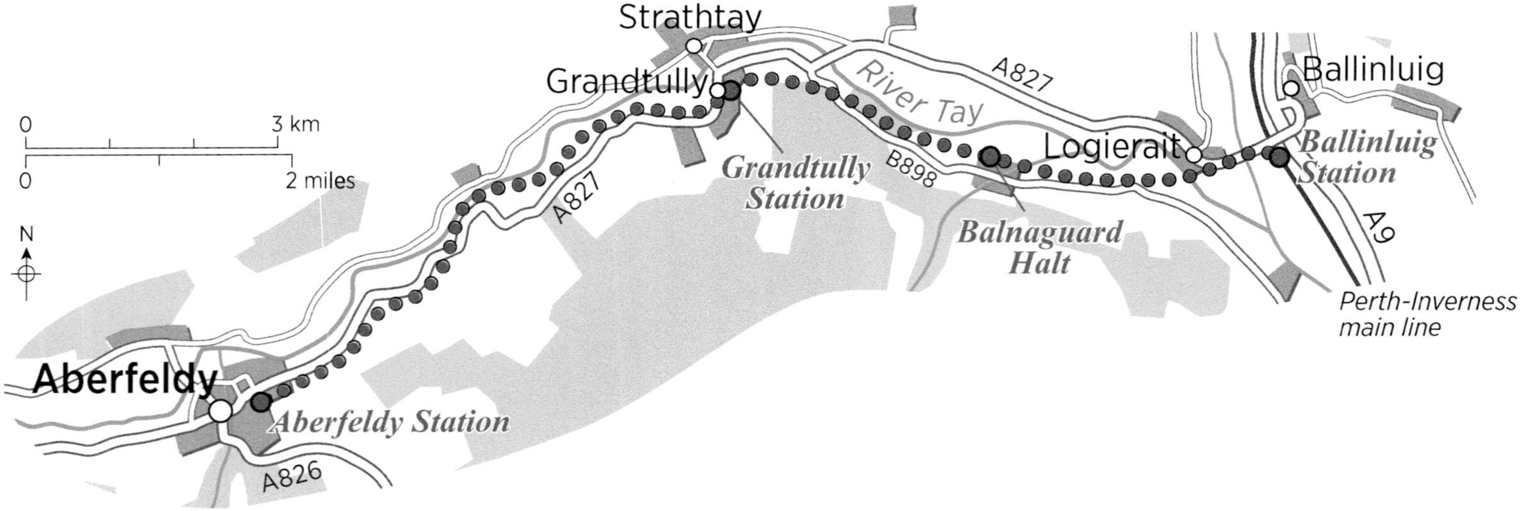
Strathtay
Grandtully
Ballinluig
A827
River Tay
Logierait
Ballinluig Station
0
3 km
0
2 miles
N
Grandtully Station
B898
A827
Balnaguard Halt
A9
Perth-Inverness main line
Aberfeldy
Aberfeldy Station
A826

Grandtully station and goods yard, now a large car park and the Scottish Canoe Association camp site. Grandtully is a noted centre for white water canoeing on the Tay and a short walk to the river may be of interest.

Grandtully to Logierait 4 miles

From Grandtully, continue on the track bed, latterly a farm track, to reach the north side of a house by the B898 and a signpost (back the way), footpath to Grandtully.

The short initial section of the track bed, on the south side of the road, is scarcely worthwhile, so walk along the road and regain the railway embankment by the stone and brick abutment of the former 20ft high bridge that spanned the narrow B898. The overgrown but discernible track bed runs parallel to the road. At one roughish section, simply use the road if preferred.

It is then a delightful way, tree-fringed and slightly overgrown, by the banks of the Tay. Reach a small bridge over a burn and a small fence, beyond which the track bed continues as a beautiful, hidden mossy way, hemmed in between river and road. Periodically used by fishermen, it is kept in good order by Sketewan Farm.

Grandtully; landscaped remains *Furnevel*

Reach a track coming in from the road. This was the scene of a 1959 derailment caused by a landslip. There were no fatalities but three passengers were injured. Continue to the outbuildings of Sketewan Farm. Other than the distant mapped bridge at Balnaguard, where the track bed resumes, all railway traces have been lost in the immediate fields.

Sketewan was the site of an Inn before the hostelry moved to Balnaguard.

On her Highland tours, Queen Victoria twice passed by the west bank of the Tay; in 1842, stopping at Sketewan whilst the horses were being changed, and again in 1866, when she complained about the newly opened, unsightly and noisy railway.

Return to the access track to regain the road which is then followed to Balnaguard, passing by the entrance to Balnaguard Glen Wildlife Reserve. On reaching the village, look out on the left for an old, broken sign, BALNAGUARD, LMS HALT PLATFORM. (This is before a wall mounted red post box.) Turn left at the sign and a grassy way leads to the stone-arched bridge and the site of the short-lived Halt, opened as late as 1935. On the east side of the bridge is a large silage dump, sweet smelling on my cold day.

A bridge carries the line over the Balnaguard Burn. Further on the track bed has been breached to give field access. A few yards beyond is a well-built stone underpass, too small for modern day use. Along this fertile upper stretch of Strathtay, it is not surprising that the gentle embankment is used for tractor access to the surrounding fields.

Later, with the river close by, the tree-fringed track bed crosses a burn (beware of a few holes) and passes close by the B898. After a minor cutting spanned by a stone-arched bridge, the sudden appearance of the grandiose Tay Viaduct where the railway spanned the river at Logierait, may come as a surprise. A ferry once operated just upstream of the bridge.

After the last train (1st May 1964), the viaduct passed into the hands of Kinnaird Estate. In 1994 the estate gifted the bridge to the local community. In 2000/2001 it was comprehensibly restored at a cost of around £400,000 and re-opened as a community-owned road bridge, used by local traffic and cyclists.

Logierait ends the walk. The bridge spanning the Tummel is no more and the junction station at Ballinluig (on the Perth-Inverness main line) which closed along with the branch, has been subsumed by the A9 dual-carriageway.

Bankfoot station building *Crawford*

Bankfoot

Almost the entire track bed of the 3-mile Bankfoot Light Railway has been lost—mostly to the re-aligned and widened A9, a little to farming. Only one too-small-to-walk section at Newmill can be identified (see general map, p. viii).

Other books for your interest

ISBN 978-184995-031-2 £16.99

- *...The narrative which accompanies each of the walks makes for entertaining reading. ...Usefully divided into different areas, with full page maps at the start of each section, this is a book to inspire walkers and to delight the armchair traveller too.* **Scottish Home & Country**
- *...The appeal in the book for me was the other walks which make up the bulk of the content. These give alternatives to Munros and Corbetts: ideal for half days and taking visitors perhaps not used to hillwalking. ... this book is ideal.* **Munro Society Newsletter**
- *...The book is ideal for suggesting options that are a bit further afield ... in the main the routes described are not more than 5 hours ... refreshment spots are suggested...***Kirkintilloch Mountaineering Club**

To order books please visit our website, www.whittlespublishing.com or call us on +44(0)1593 731333. We look forward to hearing from you.

ISBN 978-184995-162-3 £16.99

- *One of the great strengths of the book is that eight of these chapters explore the towns, countryside and Antonine Wall along the route of the canals. ... Hamish gives us many insights into local heritage.* **Canal News**

- *...a perfect companion to the canals, allowing visits to be planned and to be enjoyed to the full. ...the book is full of fascinating historical background, descriptions, stories, and practical information. It is copiously and beautifully illustrated... an ideal book for anyone who likes their walks (or cycle rides) flat rather than hilly!* **Undiscovered Scotland**

- *This excellent book, written passionately, entertainingly and informatively by Scottish canals expert, Hamish Brown, is a must for visitors to the Union and Forth and Clyde Canals. ...a very readable and lively text which captures the atmosphere of the canals perfectly...* **Towpath Talk**

ISBN 978-184995-087-9 £14.99

- *...Hamish Brown captures the atmosphere the walk generates, and many readers will identify their trip with their own personal experiences of the Way.* **The Scots Magazine**
- This is not to be missed if you're planning your own adventure down this legendary trail. **The Great Outdoors**
- *Many books have been written about the West Highland Way. Few are joyous, or enjoyable, as this one. ... The result is a book which engages you at the beginning, and keeps you engaged and amused all the way through.* **Undiscovered Scotland**

- *...it's full of fascinating snippets of information. ...it's an intriguing book... and generally adds to your knowledge of the Scottish countryside. It will be a useful and pleasurable addition to your shelf.* **Mountain Views**
- *...provides a complete list of Scottish view indicators and a useful guide of where to go for panoramas. ...reveals how a country has traditionally attracted visitors keen to know and to witness stunning scenery.* **Scottish Islands Explorer**
- *David Squire's excellent book will bring the idea of collecting view indicators to a very much wider audience than ever before. What is especially impressive is the sheer amount of background research that has so obviously gone into producing the book. ... The result is a* must buy *book for anyone who spends any time travelling Scotland.* **Undiscovered Scotland**

ISBN 978-184995-171-5 £18.99